KT-547-205

CUTTING TOOLS

CUTTING TOOLS

R. EDWARDS

THE INSTITUTE OF MATERIALS

Book 583
First published 1993 by
The Institute of Materials
1 Carlton House Terrace
London SW1Y 5DB

ISBN 0 901716 48 0

Typeset from the author's disk by
Inforum, Rowlands Castle, Hants

Printed and bound in Great Britain at
The University Press, Cambridge

Dedication

This book is dedicated to Dr. Thomas Raine who decided to employ me in 1946 in the Physical Metallurgy Group of the Research Department of the Metropolitan Vickers Electrical Company Ltd. Trafford Park Manchester.

He encouraged and made it possible for me to take my MSc and PhD degrees and gave me all my basic grounding in hardmetal. I will always be indebted to him.

R. Edwards

Acknowledgements

Grateful thanks are due to the following people and companies for their help in providing some of the information and illustrations included in this book and also for interesting and stimulating discussions on specific subjects.

John Ashley – Kennametal U.K.
Paul Bossom – De Beers
John Haddock – Hydra Tools International Plc.
Ian Hall – Cerasiv U.K.
David Hammond – Hammond & Company
Ken Foster Hudson – Foster Hudson Ltd.
Dr. David Jack – Sandvik U.K.
J.W. Lynch – International Twist Drill
John Rennie – The Rennie Tool Co. Ltd.
P.S. Thicke – Deloro Stellite Ltd.

Special thanks must be given to Mr. Horst Woehrle of Plansee Tizit, Austria. The author was associated with this company for almost 45 years until his retirement in November 1990. The majority of the photographs used in this book have been provided by Plansee Tizit and the author is most grateful to them for the help he has received.

Contents

Foreword

Any machining operation consists of three basic elements. These are: the workpiece which is to be machined, the machine on which the operation will be carried out and the tool which will be used to cut the workpiece.

In Chapter 5.1.6 workpiece materials are dealt with together with an attempt to discuss the way they machine.

The various cutting materials in use today are listed in Chapter 2 and the types of cutting tool on which they are mounted are dealt with in Chapter 3.

The main machining processes are discussed in Chapter 5 where the cutting parameters which are required for the different machining operations are described in relation to the workpiece materials involved and to the cutting material being used.

Finally in Chapter 6 a range of practical machining examples is given as an aid to the selection of appropriate cutting materials and suitable cutting conditions.

1
Introduction

This book does not deal with cutting theory. It is intended to be of help to people who know how to operate machine tools but often find it difficult to understand the recommendations for tool selection, particularly in the case of hardmetal.

The range of materials which one can use for cutting has been in a continuous state of development since the 1920s. We are now faced with the possibility of choosing from some nine classes of cutting material and then, having decided on the class of material we will use, we often have to make a further choice from within that group.

Once the cutting material has been selected cutting edge geometry comes into the equation. The rake angle, the corner radius, the clearance angle, the condition of the cutting edge itself (e.g. sharp, slight radius etc.) all play a part in optimising the performance of the chosen cutting material.

So called 'long chipping' workpiece materials (this covers almost all steels) need to have special grooves either ground or formed into the rake face of the cutting tool immediately behind the cutting edge. These grooves control the way the chip flows. Carefully designed grooves make the chip turn over on itself causing it to break into small individual pieces shaped like a figure '6' or '9'. Such pieces are ideal from the operator safety aspect and also ensure good swarf clearance and transportation. This is a vital factor with CNC machines, machining centres and flexible machining systems.

The acceptance of indexable insert tooling by the market during the 1960s has brought tremendous benefits to machine tool users. It has enabled the production of highly sophisticated chipgrooves in the indexable inserts by direct pressing techniques. It is impossible to reproduce such grooves by grinding which in any case is far too expensive. It has also resulted in the setting up of an international standard designation system for both inserts and toolholders.

One of the most important advances in cutting materials has been the development of 'coated' indexable inserts. These were first introduced in 1969. The latest generations of coated inserts bring about increases in productivity which could never have been envisaged at the time of the original development.

The toolholders on which indexable inserts are mounted have also undergone considerable change since the 60s. In particular the clamping systems have moved away from 'finger' type clamping which often interfered with chip flow, to holding by pins, or levers or by special screws which locate in a centre hole in the insert.

Automatic tool changing has become the norm in modern machine tools and there are several excellent systems in operation where the head of the tool is exchanged from a turret or magazine without any operator involvement. This enables a worn cutting edge to be changed or a new cutting geometry to be introduced to carry out another operation under programme control.

Workpiece materials are another vital factor in machining. Their physical properties and their shape are the main considerations when choosing which class of cutting material one will need. For example, on no account should one attempt to use diamond for machining steels nor should one choose a ceramic for a workpiece whose form will give rise to heavy interrupted cuts.

This book attempts to furnish the reader with information about all the points mentioned above and their relevance to turning, boring, parting, grooving, threading, milling and drilling.

2
Cutting Materials

Any machining operation which involves the removal of metal by a cutting action requires that the material used for cutting will stand up to the rigours of that cutting action.

There are three basic problems to be overcome:

a) The wear which takes place at the cutting edge.
b) The heat generated by the energy required to remove material from the workpiece.
c) The shock involved in the cutting action.

The main properties which any cutting material must possess in order to carry out its function are therefore:

a) Hardness to combat the wearing action.
b) Hot strength to overcome the heat involved.
c) Sufficient toughness to withstand any interruptions or vibration occurring during the machining process.

The following materials are those generally used for cutting:

High Speed Steels
Stellite
Hardmetals
Cermets
Sialons
Ceramics
Silicon Nitride
Cubic Boron Nitride
Diamond (Man Made & Natural)

Except for Hardmetals they are listed in order of hardness. Hardmetals cover a wide range of hardness and overlap cermets and sialons at their harder end.

In general, increasing hardness brings with it a reduction in toughness and so those materials in the higher hardness region of the list will fail by breakage if used for heavy cuts, particularly with workpieces which have holes or slots in them which give rise to interruptions in the cut.

Sialons and silicon nitride are also regarded as ceramics. There are two generally recognised groups of ceramics and they fall into the group known as silicon-based ceramics whilst the other group is known as the aluminium oxide-based ceramics.

2.1 HIGH SPEED STEELS

High speed steels have the lowest hardness and the highest toughness of the cutting materials in general use. Their major disadvantage is that their hardness is brought about by a heat treatment process so they are not naturally hard. If the temperature of the cutting edge rises to around 600°C then the High Speed Steel will soften and the edge will fail. For this reason they are limited to comparatively low cutting speeds up to a maximum of the order of 50 m min^{-1}.

In turning they are mainly used as circular, or dove tailed, form tools on so called 'automatic screw machines'. These machines are found in establishments producing high volume parts directly from bar or tube. Many of them are multi-spindle machines where cutting is going on at most of the stations at the same time. The form tools take a broad cut and because the machines lack rigidity by the nature of their design this broad cut must be comparatively light. This in turn calls for high rake angles on the tools (usually a minimum of 10°) which results in a weaker geometry at the cutting edge. The use of harder cutting materials would require reduced rake angles and higher cutting speeds but the multi-spindle autos are not rigid enough to work continuously under such conditions and high machine maintenance costs together with frequent tool breakage would result.

Probably less than 10% of all turning applications are carried out using high speed steel as the cutting material. Their major area of application is drilling. At least 80% of all drilling is done with high speed steel. It is ideally suited for most of the machines in use today which have insufficient power and lack the rigidity so necessary for drilling with hardmetals.

The second most important application area for high speed steel is milling. Solid high speed steel end mills, slot drills and router cutters form a large market and together with high speed steel face and corner milling cutters up to 75mm and 100 mm in diameter they make up about 40% of the total milling cutter market.

Although the development of CNC machines and machining centres equipped with robust rotating spindles and the introduction of stiffer drilling machines and milling machines with more power available has aided the increased use of hardmetal, high speed steel is still likely to be the predominant material for drilling and a widely used material for milling.

Another important development which has enabled high speed steel to cut at higher speeds has been the utilisation of a very thin titanium nitride coating on the surface of the tool. This is particularly so in the case of drills where both increased feed rates and cutting speeds have resulted. This TiN coating, which is gold in colour, is about 3 micrometres thick and is extremely hard and stable. It is applied by a process known as Physical Vapour Deposition (PVD) whereby the high speed steel base material does not reach a temperature greater than 500°C and thus its hardness is unaffected. This coating technique is dealt with fully in Chapter 4.

Coating is ideally suited to tooling which is not reground when the cutting edge is worn. Form Tool Systems, a UK company, have perfected an excellent clamped TiN coated insert system for circular and dove tailed form tools and also an indexable coated insert system for parting and grooving tools (Fig. 1). The substrate of the inserts is conventional high speed steel.

More recently Plansee TIZIT have developed a range of high speed steel TiN coated indexable insert tooling for turning and parting. The substrate of the inserts is powder metallurgy high speed steel and this has enabled them to press in specially designed chip control grooves which optimise the cutting geometry of the inserts (Fig. 2).

Three types of high speed steel are available. The first uses tungsten as its major alloying element and in the UK is known as the 'T' series. The second type contains molybdenum and considerably less tungsten is present. This is known as the 'M' series of alloys. The third type contains cobalt and can be either a T or an M series of material.

The T series without cobalt are not quite so tough as the M series but their heat treatment is easier to carry out. The M series are more widely used, especially with drills and end mills. The introduction of cobalt increases hot hardness and wear resistance but reduces the toughness. High speed steels containing cobalt appear to be more advantageous when machining steels with a hardness over 275 Brinell. A British Standard exists (B.S. 4659).

Fig. 1 Indexable Insert High Speed Steel Tooling

Fig. 2 High Speed Steel Indexable Inserts

The hardness of high speed steels after they have been heat treated is usually quoted in Rockwell C units and generally falls within the range 62 to 68 Rc. However in this book the hardness of all the other cutting materials is quoted in Vickers Diamond Hardness so for comparison purposes we can say that high speed steel lies in an approximate range of 800 to 900 VDH.

The most popular alloy for producing drills is M2 and this is also a favourite for the production of taps. It is extremely unlikely that any of the other cutting materials will succeed as a basis for standard taps. The harder T42, ca. 1000 VDH, is used when abrasion resistance of the cutting edge is the vital factor. If hot strength is the main requirement such as in the machining of heat resisting alloys then M42 is used. On the continent of Europe M35 is the choice for similar applications.

M42 is also the ideal substrate for coated inserts made from conventional high speed steel whereas M35 is currently the preferred material for powder metallurgy High speed steel with a TiN coating.

Figure 3. shows the structure of wrought M3 high speed steel heat treated in vacuum, the magnification is × 500. Figure 4. shows an M35

Fig. 3 M3 High Speed Steel Heat Treated × 500

Fig. 4 Powder Metallurgy M35 Heat Treated × 500

high speed steel at the same magnification, made by powder metallurgy and similarly heat treated.

Summarising

Main areas of application: Drilling, End Mills, Solid Milling Cutters, Slot Drills, Circular and Dove Tail Form Tools, Taps, Reamers, Broaches, Hobs, Butt Weld Turning Tools, Regrindable Tool Bits for smaller and lower powered lathes. High speed steel is restricted to comparatively low cutting speeds. Higher speeds will cause the temperature of the cutting edge to rise above the softening point.

2.2 STELLITE

Stellite is the trade name for a cobalt-based alloy which is naturally hard and does not require heat treatment to attain its cutting properties.

Originally two compositions of stellite were supplied for metal cutting. Now only one grade is offered and is known as Stellite Alloy No. 100. This is a cobalt alloy containing chromium, tungsten and carbon. It is produced by melting and casting and is as hard as the hardest high speed steels but its hot hardness at dull red heat is 535 VDH compared with 175 VDH for high speed steel. It is mainly used for turning operations and is supplied as solid tool bits and as turning tools which are tipped with the Stellite alloy. The cutting geometry is ground into the tool and once the cutting edge is worn it is reground to bring it back to new condition.

Stellite tools are used to cut surfaces which are extremely difficult to machine with hardmetals and where the cutting edges of hardmetal would be liable to fracture (hardmetals are described in 2.3). A typical example is the machining of welds. Welds tend to be hard and have inclusions in their surfaces. They are uneven and give rise to interrupted cutting.

Stellite is tough enough to cope with these conditions even with positive rake geometry. The range of cutting speeds in which it will perform satisfactorily is lower than that for hardmetals but a little higher than that for high speed steels.

Properties

Stellite Alloy No. 100:

Composition – 34% Cr, 19% W, 2% C, balance Co.
Hardness – ca. 950 VDH.
Hot Hardness – 535 VDH at 700° C
Density – 8.75 g cm^{-3}

A photomicrograph of Alloy No. 100 is shown in Figure 5 at a magnification × 100. The structure is typical of a cast material.

Summarising

Stellite will perform on heavy cutting operations at medium to low speeds. It is not one of the important cutting materials and has a narrow, specialised field of application.

2.3 HARDMETALS

This family of alloys is the hard core of all the cutting materials in use today. There is no international standard based on composition and

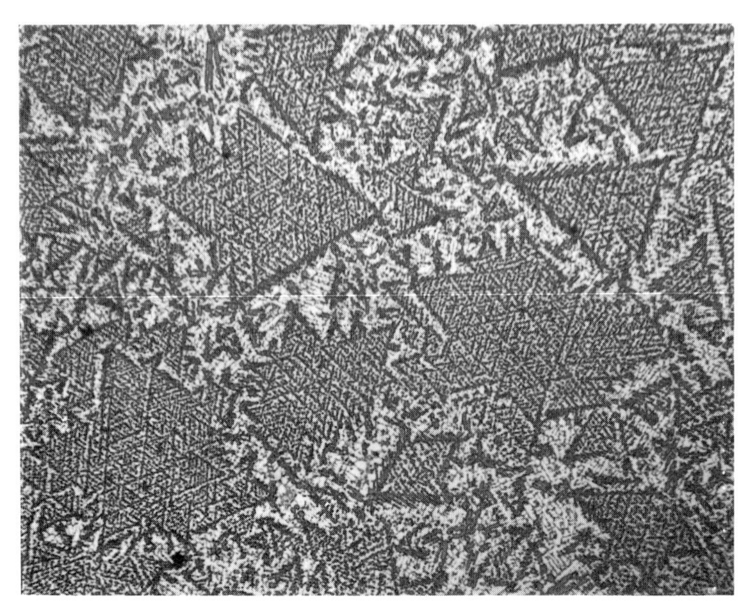

Fig. 5 Stellite Alloy No. 100 × 100

mechanical properties for hardmetals. There is, however, an ISO standard for machining applications. Hardmetal manufacturers then nominate alloys from their range which they recommend to carry out the ISO applications. The alloys are usually called 'grades'.

This application standard is ISO R513 and it classifies workpiece materials into three major groups. Each group is given a letter and a colour to identify it. The cast irons and non-ferrous metals applications are given the letter K and their colour is red. The steels group has the letter P and is coloured blue. The third group covers more difficult materials like heat resisting alloys and is given the letter M and is coloured Yellow.

The groups are then sub-divided into the types of application involved. Lighter, finishing cuts are at the top of the group and heavy roughing cuts at the bottom. Each type of application is given a number. The smaller numbers relate to lighter cuts and the larger numbers identify the roughing applications. Thus fine finish turning of a mild steel cylinder with no interruptions would be termed a P05 application whilst planing a cast iron lathe bed with interruptions and sand inclusions would be termed a K40 application.

The grades which any two hardmetal manufacturers nominate to carry out a P05 application will almost certainly not be identical in composition but they are likely to be near to one another and their properties will be similar. This comment applies all through the range.

Although this system does not classify competitors' cutting materials as direct equivalents nevertheless it has to be said that, by and large, it works.

Prior to the introduction of coatings in 1969 two groups of hardmetal existed for machining purposes. Both these groups are still used but they have been joined by a third group of coated hardmetals which can perform many of the tasks previously carried out by the original hardmetals. Coated hardmetals will be described fully in Chapter 4.

The simplest hardmetals are the first group and are composed of tungsten carbide (WC) bonded by cobalt (Co). Tungsten carbide has a hardness in excess of 2000 VDH whilst cobalt has a hardness only 10% that of tungsten carbide. Pure WC is comparatively brittle and Co is tough. A combination of these two materials results in a compromise between wear resistance and shock resistance according to the amount of Co binder used. The quantity of Co contained in a hardmetal is usually reported in weight percent. Because the density of WC is almost twice that of Co the volume of binder material is considerably greater than would appear from the quoted Co percentage.

Two factors affect the cutting properties of a simple WC-Co hardmetal. They are:

a) The cobalt content
b) The grain size of the tungsten carbide

Increasing the Co content increases the toughness of a hardmetal but reduces its hardness and therefore its wear resistance. Coarser grain WC is better for shock resistance and for a given Co content reduces the hardness of an alloy compared with finer grains.

Conversely, reducing the Co content reduces the toughness and increases the wear resistance by increasing the hardness of a hardmetal. Fine grain WC also increases the hardness and therefore the wear resistance for a given Co content.

The useful range of Co content for cutting purposes in weight percent is from around 5% to 12%. Grain sizes of WC go from around 0.5 micrometres to 5 micrometres. The hardness span of these alloys ranges from 1250 VDH to 1800 VDH.

The structure of a 6% Co 94% WC hardmetal is shown in Figure 6. This photomicrograph is taken at a magnification × 1500. The grey

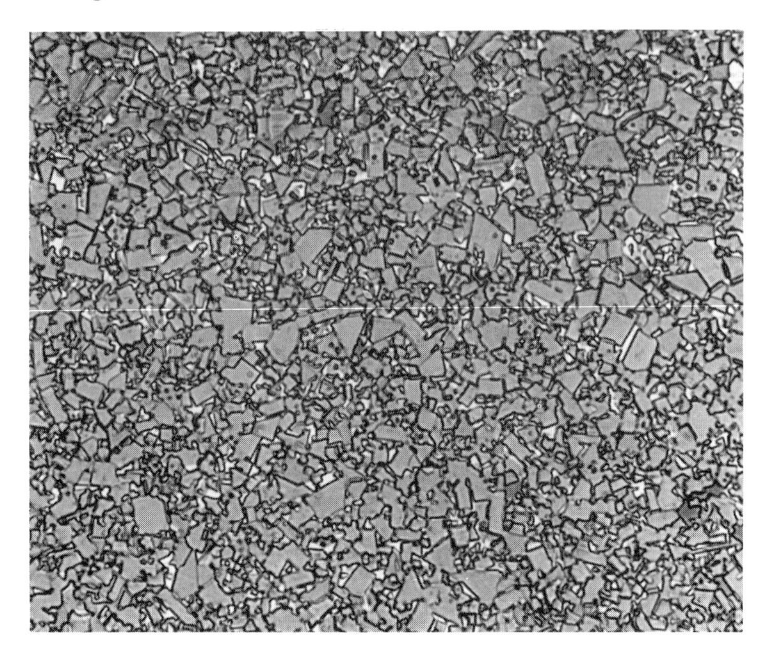

Fig. 6 6% Co 94% WC hardmetal, medium grain × 1500

angular grains are the WC and the white background is the Co binder metal. In this case the grain size of the WC is an average of about 1.5 micrometres and is termed a medium grain size. In the USA and in other parts of the world, hardmetals are known as 'cemented carbides' and the photomicrograph clearly shows that this is a very descriptive term for this class of cutting material.

The effect of a change in grain size but keeping the same chemical composition is illustrated × 1500 in Figure 7. In this case the grain size is 0.5 to 0.7 micrometres and is termed ultra-fine. This grain size is now at the limit of resolution of the optical microscope so it is almost impossible to discern the grain boundaries and shape of the WC grains. The surface area of these very fine grains is very large compared with the medium grains and so the Co is very thinly spread over this surface area and is barely visible. If the structure is examined under an electron microscope where a higher magnification can be used – say × 4000 – the WC grains appear very much as in Figure 6.

The hardness of the medium grain hardmetal containing 6% Co is ca. 1600 VDH and its density is 14.9 g cm^{-3}. The ultra-fine grain hardmetal has a hardness of 1800 VDH with the same density.

Fig. 7 6% Co 94% WC hardmetal, ultra fine grain × 1500

The medium grain material represents the compromise between hardness and toughness for this particular composition and is a very popular choice of cutting material for cast irons, austenitic stainless steels and non-ferrous metals. However, when cutting ferritic steels the problem of crater arises and the plain Co-WC alloys are no longer able to perform.

A cutting speed of 50 metres per minute is at the low end of the range of speeds typically used for machining with hardmetals. Even at this speed, when cutting steel, the temperature at the interface between the chip and the tool tip is well over 1000° C. At these temperatures iron is able to absorb tungsten carbide by a mechanism known as solid solution. The way in which this operates when cutting ferritic steels with a Co-WC alloy is that a crater is formed immediately behind the cutting edge. Metallographic examination of the chips reveals grains of WC which have been removed from the tool tip. The higher the cutting speed, the higher the temperature, the more rapid is the cratering effect and breakdown of the cutting edge occurs in a very short time.

In order to be able to machine ferritic steels it is necessary to make the hardmetal resistant to cratering. This is done by adding titanium carbide (TiC) to the basic Co-WC alloys. These materials form the second group of hardmetals used for machining.

TiC has an extremely low solubility in iron and therefore as the chip flows over a cutting tool tip containing grains of TiC they act as a barrier and deter the cratering action. The hardness of TiC is even harder than WC and therefore wear resistance is maintained.

The amount of TiC added varies from about 5% to 25% by weight. Its density is only 4.9 g cm^{-3} compared with 15.7 for WC and so much more TiC is apparent in the structure by volume than would be imagined from the weight percentage. The proportion of TiC added depends on the cutting speed the hardmetal is required to perform at. Finishing operations need to be carried out at higher speeds for economic metal removal. High speeds will cause cutting temperatures to increase and cratering will be more pronounced. To counteract this a high TiC addition is made. Adding TiC tends to reduce the toughness of the alloy but with finishing operations the cutting is very light and the hardmetals containing up to 25% TiC by weight are tough enough to perform satisfactorily.

At the other end of the scale heavy roughing operations are usually carried out at lower speeds and so the cutting temperature is lower and cratering is reduced therefore less TiC is needed. The smaller amount of TiC does not adversely affect the toughness of the hardmetal.

Tantalum carbide (TaC) has also been added to these TiC containing grades since the mid 1950s. It increases the hot hardness of the alloy and this helps to prevent plastic deformation of the cutting edge. Because TaC is expensive it is often diluted with up to 50% niobium carbide (NbC) without detracting from the performance of the alloy.

A photomicrograph of one of this group of hardmetals is shown in Figure 8 at a magnification × 1500. The composition of this alloy is 8.5% Co, 71.5% WC, 9% TiC and 11% TaNbC. Its hardness is 1575 VDH and its density is 12.4 gms. per cc. The angular, lighter grains are the WC and the white background is the cobalt binder. The darker more rounded grains are what is termed a 'TiC mixed crystal' or solid solution of TaNbC + WC in TiC.

All the family of hardmetals is produced by a powder metallurgy process. The basic Co-WC alloys are made by mixing cobalt and tungsten carbide powders, pressing the mixture into shapes and then sintering these shapes. When TiC is to be included in the alloy it is best to add it as a powder which is a solid solution of WC (and TaNbC where needed) in TiC. The resulting alloy is tougher than if TiC is added as pure powder.

Fig. 8 Crater Resistant TiC containing Hardmetal × 1500

Reverting back to the ISO application standard, the grades which are used for the K applications i.e. to machine cast irons, austenitic stainless steels and non-ferrous metals are the plain Co-WC hardmetals. The K30 and K40 applications require toughness and therefore need a hardmetal with a high Co content to withstand the shock. The grain size must be at least medium and tending to coarse for the really tough applications. The very light finishing operations, K01, present no problems of toughness and so the hardest, most wear resistant, plain Co-WC grades are used i.e. 5% Co and fine grain WC.

The P applications need hardmetals containing TiC to combat the problem of cratering. The heavy, interrupted, roughing operations need a high Co content and a medium to coarse grain size of WC to withstand the shock during cutting. This will result in a hardness in the region of 1400 VDH.

A typical hardmetal for finishing operations will have a low Co content e.g. 6% to 7%, a high TiC content of around 20% and TaNbC of the order of 10%. The hardness of such an alloy will be 1700 VDH.

The range of hardmetal alloys used for the M applications is much narrower. Co contents are from 6% to 9%, TiC from 4% to 8% and TaNbC from 5% to 10%. They lie in a hardness band of the order of 1450 to 1650 VDH.

The information given above relates to normal hardmetals. Over the last twenty years remarkable improvements in their cutting performances have been achieved by applying very thin coatings of TiC, TiN, TiCN, Al_2O_3 etc. This is an extremely important subject in metal cutting and so a separate chapter is devoted to coating.

Summarising

Hardmetals cover a very wide band of machining applications. It is estimated that some 70% of all turning tasks are done using hardmetal tooling. A range of compositions is available and each alloy is tailor made to provide the properties needed to perform the special requirements of an application e.g. high hardness for finishing or good toughness for roughing. Coated hardmetals, in the form of indexable inserts, enable very high productivity levels to be achieved (see Chapter 4).

2.4 CERMETS

Cermets are carbonitride based materials. They have TiCN as the major hard phase which is held together by a softer binder alloy of Co and/or Ni. The grain size of the TiCN is usually in the range 0.5 to 2 micrometres and an electron microscope photograph which illustrates the structure of a cermet is given in Figure 9. In this case the average grain size of the TiCN is of the order of 2 micrometres. Cermets have a density in the region of 6 gms. per cc.

Each manufacturer has his own compositions and many include carbides such as Mo_2C, WC and TaC. The hardness of these cermets is around 1600 VDH.

This group of cutting materials has made a major penetration into the Japanese market where approximately 25% of all indexable inserts are cermets. In Europe this figure is a maximum of 3% to 4% and does not appear to be growing to any great extent at the time of writing this book. There seems to be no single special reason why Japan has such a high usage but factors which are put forward are:

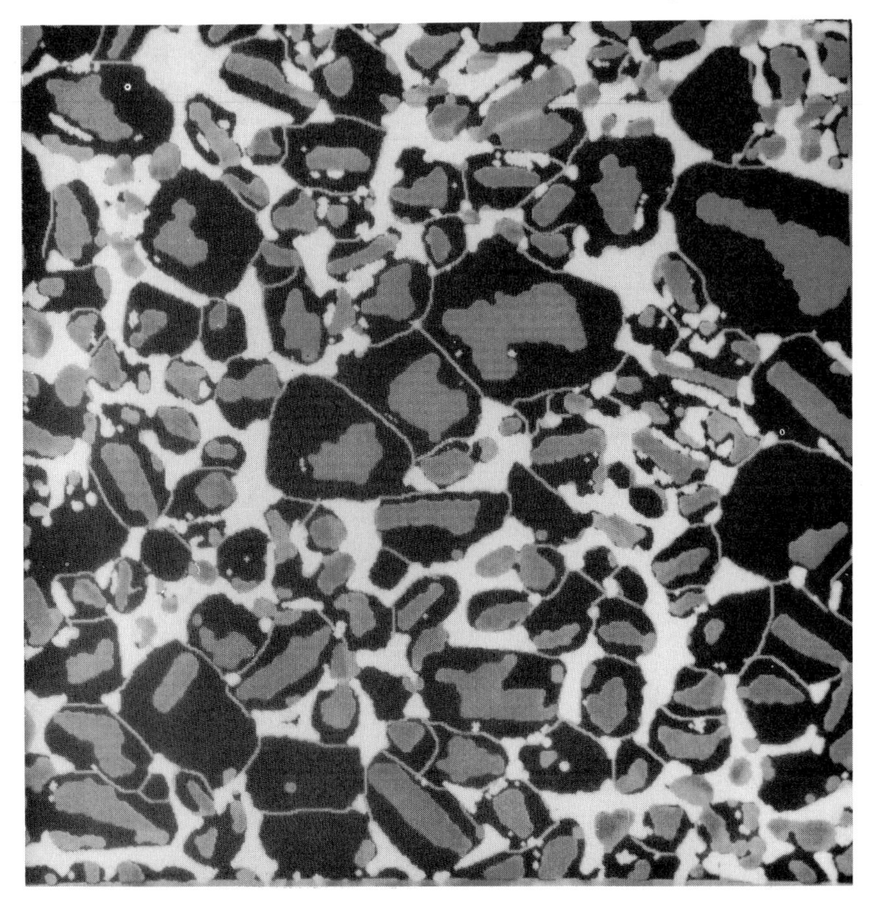

Fig. 9 Microstructure of a TiCN based Cermet × 5000

a) Attempts to replace tungsten, which had become a strategic metal in the 1950s and 1960s, accelerated the development of cermets as cutting materials in Japan.
b) Japanese machine tool technology is very advanced – stiffer machines with more power are a prerequisite for applying cermets.
c) Development of coated hardmetals in Japan was much slower than in Europe and USA and cermets are used in Japan in applications where coated hardmetal would be the normal cutting material in the West.

Although their hardness is similar to that of hardmetals they have a lower susceptibility to diffusion, favourable frictional behaviour and low susceptibility to wear. These factors make it possible to cut clean metal at higher speeds than would normally be used for hardmetal. Their major area of application is on finish turning and semi-finish turning of steels. Excellent results can also be obtained with light milling cuts on difficult to machine steels such as stainless steel.

Summarising

Cermets are capable of machining at speeds which exceed those normally applicable to hardmetals. They perform well with medium to light cutting in both turning and milling applications on steel workpieces.

2.5 CERAMICS

2.5.1 SIALONS

These cutting materials can be classed as ceramics based on silicon nitride. Powders of silica (SiO_2), alumina (Al_2O_3) and silicon nitride (Si_3N_4) are mixed with a small addition of yttria (Y_2O_3) then cold pressed and sintered. The Y_2O_3 aids sintering, MgO is also used as an alternative.

During sintering the silica reacts with the alumina and the yttria to form a liquid. The silicon nitride then reacts with this liquid to form sialon (silicon aluminium oxynitride) and, on cooling, a glass.

Their microstructure is composed of grains of the crystalline nitride phase in a glassy matrix. Figure 10 shows the structure of a sialon × 5000 which consists of a mixture of beta sialon and intergranular glass/crystalline phases. The hardness of sialons is ca. 1700 VDH and their density is around 3.3 g cm^{-3}. They have a very low coefficient of thermal expansion which reduces the stresses set up between the hotter and cooler parts of a cutting insert and so their thermal shock resistance is excellent.

Sialons retain their hardness better than alumina at temperatures of 800° to 1000°C. This is one of their outstanding properties. However, they do not have the same toughness as a hardmetal of equal hardness. Their properties make them very suitable for machining heat resisting alloys although whisker reinforced ceramics are making inroads in this area.

Sialons cannot be used for general steel machining at high speeds because of rapid solution wear. They can perform very effectively when

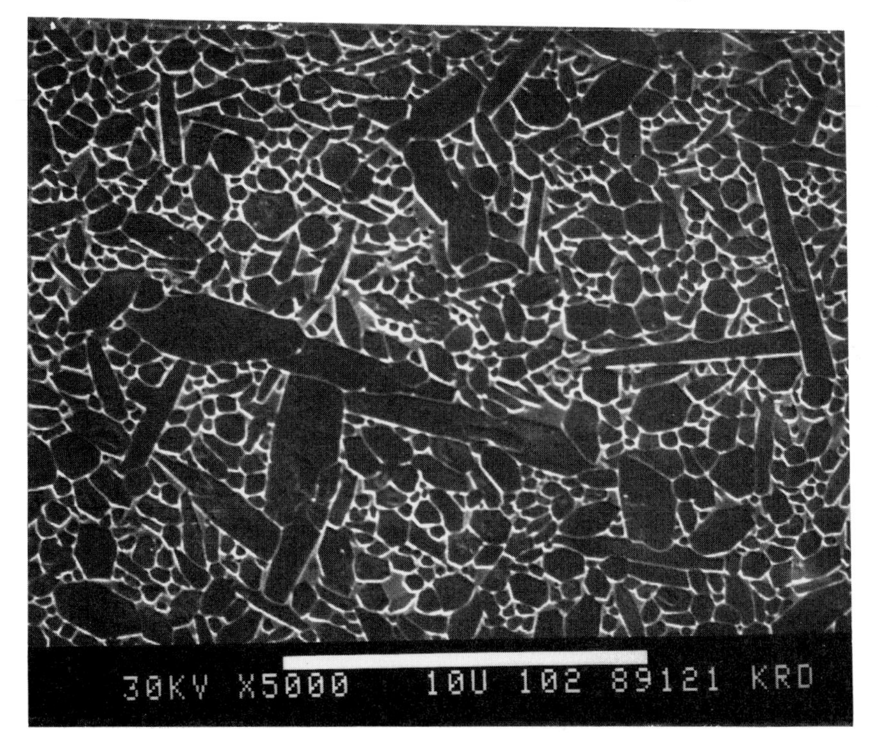

Fig. 10 Microstructure of a Sialon × 5000

machining hardened die steels. Cast irons can be machined at higher speeds than with uncoated or coated hardmetals and higher feeds can be used than with conventional alumina based ceramics.

Summarising

The properties of sialons make them suitable for machining heat resisting alloys. They perform well on cast irons at high cutting speeds but are not suitable for general steel machining.

2.5.2 ALUMINIUM OXIDE-BASED CERAMICS

In metal cutting the prime factor that has to be coped with is the heat generated during the machining operation. At around 800 °C Al_2O_3 ceramics begin to show better mechanical properties than hardmetals, particularly in compression. Below 800° C hardmetals have superior strength compared with ceramics.

Ceramics are especially good for machining grey cast iron in large series production. The automotive industry is the major example of this. In Germany ceramic cutting tools are used for machining many of the brake discs, brake drums and flywheels which are produced there.

Three types of Al_2O_3 ceramic are described in this book. The simplest ceramic is white in colour and comprises alumina with additions of from 2% to 5% of zirconia (Al_2O_3 + ZrO_2). Zirconia increases the fracture toughness without affecting wear resistance. This ceramic has a low thermal conductivity which makes it susceptible to thermal shock and so the use of coolant should be avoided. It is made from powders which are hot pressed and this presents a limitation on the shapes which are able to be produced. Its density is around 4 g cm^{-3} depending on the amount of ZrO_2 present. Hardness values reported vary but certainly lie right at the top end of the hardmetal range. This alumina/zirconia white ceramic can be regarded as the 'tough' grade for cast iron and steel machining and in cases where ceramics can be applied will do the heavier work.

The hardness of alumina can be increased by the addition of between 30% and 40% of TiC or TiN (TiC seems to be the most popular addition). Such additions push up the room temperature hardness to some 200 VDH higher than the white ceramic and have a similar effect on the hot hardness, they do, however, reduce the toughness. The increased hardness makes it more suitable for finishing operations and for machining harder steels. The colour of this second type of ceramic is black or dark brown depending on whether TiC or TiN has been added.

The third category of ceramic based on alumina/zirconia includes silicon carbide (SiC) 'whiskers', 25% or more, which act to reinforce its structure and increase its toughness. These whisker reinforced ceramics are particularly recommended for semi finish machining and finish machining of nickel based superalloys at high cutting speeds. Their hardness is of the order of 2000 VDH. Figure 11 is a photomicrograph showing the structure of a whisker reinforced ceramic. It consists of silicon carbide whiskers in an alumina/zirconia matrix. The whiskers vary in length from 10 to 50 micrometres and have cross section of 0.5 micrometres.

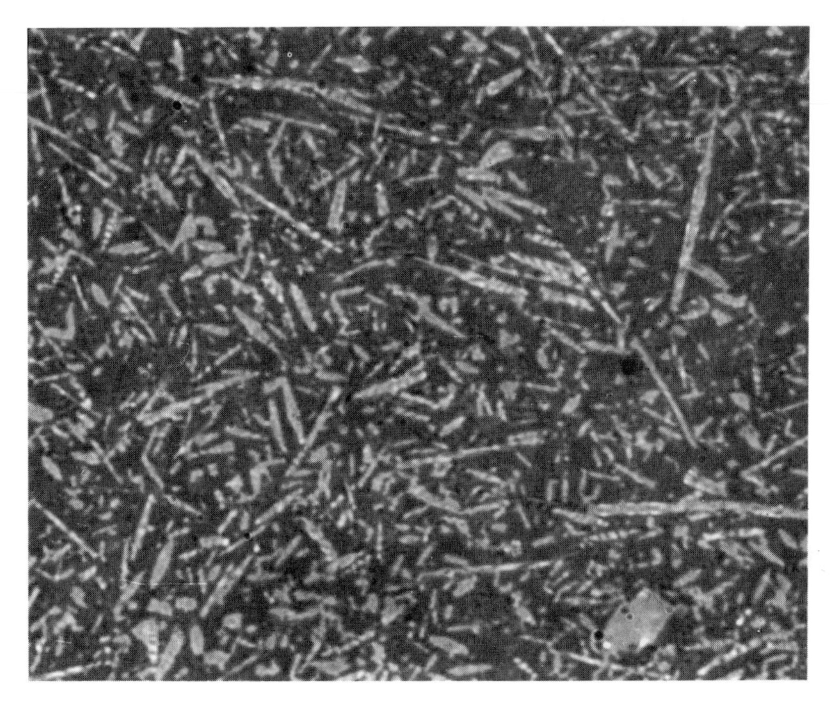

Fig. 11 Microstructure of a Whisker Reinforced Ceramic × 1500

Summarising

The alumina based ceramics have a higher hot hardness than hardmetals and therefore can operate at higher speeds without plastic deformation of the cutting edge occurring. Their higher hardness makes them more sensitive to shock and so their range of application is generally confined to clean cutting, semi finishing and finishing machining operations. On hard materials they can produce surface finishes which are normally obtainable only by grinding – hard steel rolls is a good example.

2.5.3 SILICON NITRIDE

Silicon nitride (Si_3N_4) is a ceramic which is used as a cutting material. It is made into shapes by a powder metallurgy process but does not sinter readily to full density. Some manufacturers make small additions to assist sintering but a hot pressing technique has to be used to achieve good strength. The hardness of this cutting material is ca. 1800 VDH and its density is 3.2 g cm^{-3}. The grain size is in the range 2 to 3 micrometres.

It has good wear resistance and cutting edge strength, it also has high resistance to thermal shock. Its main application field is as a roughing grade for turning and milling cast iron. However, Si_3N_4 is worse than WC for solution wear and so it is totally unsuitable for machining steels.

The good resistance to thermal shock permits machining either with or without coolant.

Summarising

Silicon nitride has good toughness which permits rough turning of grey cast iron with interrupted cuts and milling of grey cast iron at high feed rates per tooth.

It should not be used to machine steels.

2.6 CUBIC BORON NITRIDE (CBN)

Cubic boron nitride (CBN) is not a naturally occurring compound. Normal boron nitride has a hexagonal crystal structure but if it is heated to a temperature around 1400°C and a pressure of the order of 60 kilobars is then applied the hexagonal crystals are converted into a cubic structure which is the same as that of diamond and is extremely hard – of the order of 4000 VDH. A small amount of catalyst is also involved to assist the conversion.

CBN is polycrystalline and is used as a cutting material when hardmetal becomes limited in the cutting speeds that can be employed. This applies to hard workpiece materials such as high speed steel, tool steels, case hardened steels, chilled cast iron, stellite etc. It offers no advantages and does not perform well on soft steels, inconel and nimonics and austenitic stainless steel.

CBN is offered either as solid indexable inserts or as inserts consisting of an upper face of CBN laid onto a hardmetal base or thirdly as a piece of CBN brazed onto a corner of a hardmetal indexable insert. Solid indexable inserts are particularly suitable for heavier roughing work, especially for machining rolls. Round inserts can be used to machine hardmetal rolls containing 15% Co and over.

A second CBN cutting material has been developed which incorporates other hard compounds such as titanium carbide or titanium nitride. This reduces the thermal conductivity of this material to about 50% that of 'pure' CBN.

Figures 12 and 13 show the microstructures × 1000 of 'pure' CBN and of TiC diluted CBN. They are known as high CBN content and low CBN content cutting materials respectively.

The low CBN content cutting material has a greater resistance to wear than the high CBN content material under light cutting conditions and yet it has a somewhat reduced hardness.

When machining hardened components with CBN the combination of negative rake and high cutting speeds generates heat to deform the workpiece material in the shear zone. When the depth of cut is reduced less heat is generated and the high thermal conductivity of CBN rapidly takes the heat away from the cutting zone. This results in less softening and makes it more difficult for the tool to deform and shear the workpiece efficiently. At these lower depths of cut under the same conditions the much reduced thermal conductivity of low CBN content material restricts heat transfer and concentrates it in the shear zone thus producing favourable softening and reducing wear on the cutting edge.

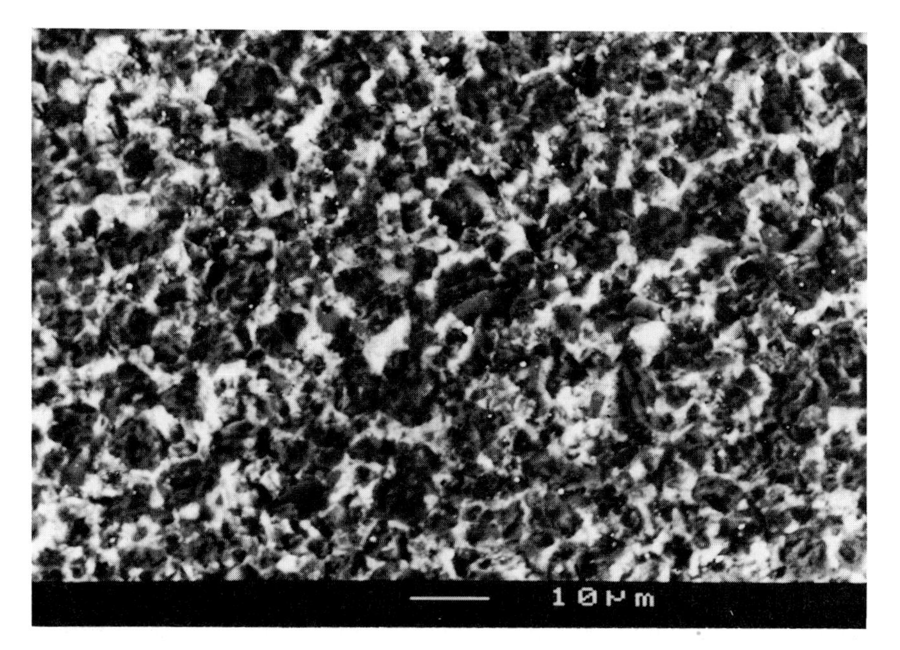

Fig. 12 'Pure' Cubic Boron Nitride × 1000

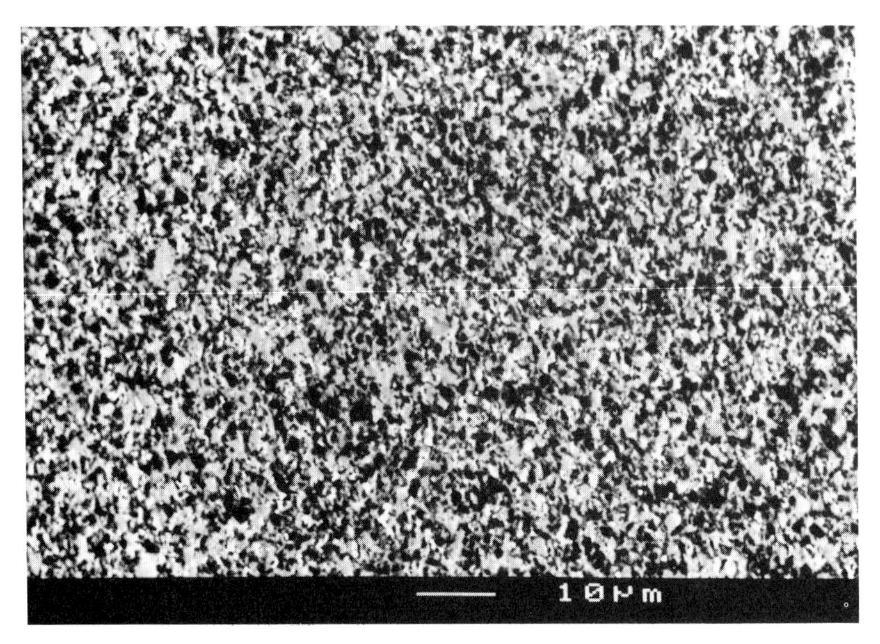

Fig. 13 Cubic Boron Nitride with TiC added × 1000

The low CBN content material is supplied as a brazed tip on the corner of an indexable insert.

Summary

With the exception of diamond, cubic boron nitride is the hardest of the cutting materials in use today. It is outstanding when machining hard materials but offers no advantage on soft workpiece materials.

2.7 POLYCRYSTALLINE DIAMOND (PCD)

Since 1958 it has been possible to produce diamonds synthetically. These 'man made' diamonds are manufactured by taking carbon and applying extremely high pressure at a temperature in the region of 1500°C. A pressure of 60 kilobars is applied to the heated graphite which, in the presence of a catalyst, is converted to diamond. Hardmetal dies are used for this purpose and are necessary to transmit the pressure. The particles of diamond are recovered from the capsule by acid dissolution.

These diamond particles are then fused together into a mass of many crystals – polycrystalline diamond – and this makes the manufacture of comparatively large pieces into a commercial proposition.

This PCD material is fused onto a backing of hardmetal and discs of up to 34 mm in diameter can be made. These discs are cut into smaller pieces which are then used to 'tip' cutting inserts. They are brazed at temperatures which must not exceed 800°C or the diamond will begin to revert to graphite.

The hardness of polycrystalline diamond approaches that of natural diamond but is not equal to it.

The structure of a PCD material is shown in Figure 14 × 1000. In practice three grain sizes of PCD are available and these are termed fine, medium and coarse. The fine grain PCD is slightly less shock resistant than the coarse material but has somewhat higher wear resistance and vice versa. The medium grain is the compromise between the other two. The PCD supplier can be expected to supply the most suitable material for each application.

PCD is suited for machining soft abrasive non-ferrous materials and especially at very high cutting speeds. It is the hardest material we know of and has superior abrasion resistance to any other cutting material. Free machining aluminium alloys, high silicon aluminium alloys, non-

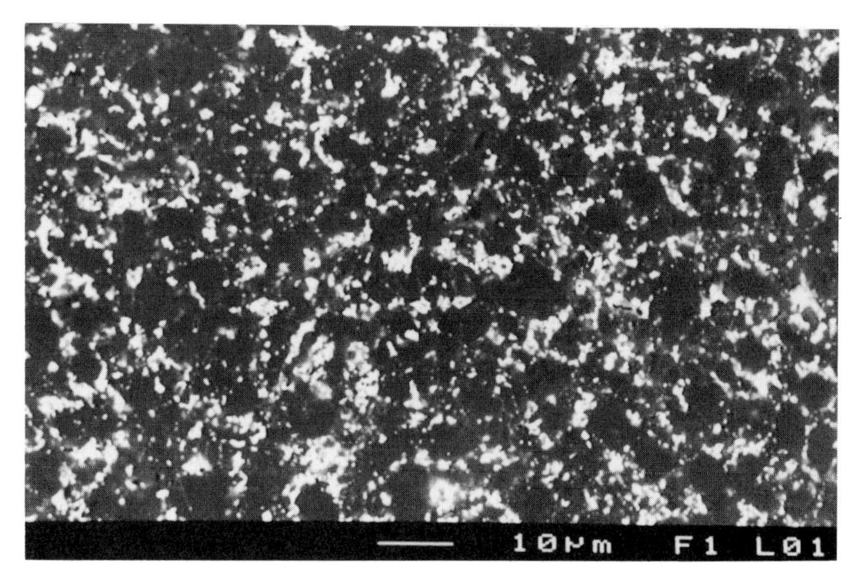

Fig. 14 Polycrystalline Diamond × 1000

ferrous metals such as copper, zinc and brass alloys or non-metallic materials are all ideal workpiece materials to be machined with PCD. It should not be used to machine steels. If one attempts to machine ferrous materials with diamond a reaction takes place between the workpiece and the diamond cutting edge causing rapid breakdown to occur.

A very special application is the machining of hardmetal. Hardmetal dies can be turned using PCD and this shows a considerable saving in time compared with grinding.

There are some instances where natural diamond gives a better result then PCD. These are cases where surface finish is absolutely critical e.g. metal mirrors, contact lenses etc. A natural diamond is an individual crystal and when a cutting edge is produced from it then that edge is one perfect line with no junctions along it. A PCD material gives a cutting edge which is made up of a number of crystals joined together. Each joint will leave a minute mark on the workpiece which can be optically unacceptable. Natural diamond leaves no such marks.

Summary

PCD is a synthetically produced diamond cutting material which is extremely hard and therefore has superb wear resistance. It is able to cut at very high speeds on soft non-ferrous workpiece materials. It is susceptible to chipping if subject to impact. It should not be used on steels and cast iron. It is increasingly used in the automobile industry on aluminium alloys where it achieves excellent surface finishes at very high cutting speeds. It can be used for machining hardmetal under the right conditions.

3
Brazed Tools and Indexable Inserts

For a variety of reasons, not the least of which is expense, it is not feasible to make turning tools and multi-toothed milling cutters from solid hardmetal unless they are very small. In these cases the body of the tool is made from steel and a hardmetal piece is attached to the cutting point, or points, of the tool.

Two methods of attachment are used:

The first is to braze, or hard solder, the hardmetal pieces to the steel shank or milling cutter body.

The second is to securely clamp the hardmetal pieces to the steel shanks or milling cutter bodies.

3.1 BRAZED TOOLS

Brazed tools present disadvantages in machining operations and some of these are outlined below.

When hardmetal is brazed to steel considerable stresses are set up resulting from the differing coefficients of thermal expansion of hardmetal and steel. The cutting edge of a brazed tool is always in tension due to this residual stress.

When brazed tools have become worn they are removed from the machine and reground. Once the cutting edge has been resharpened the tool is refitted. It is then necessary to reset the location of the cutting edge to its original position. At least one component is then machined to ensure that the resetting has been carried out correctly, if not the necessary adjustment is made.

In order to keep the time the machine tool is out of action as short as possible at least one other brazed tool must be available ready ground and kept at hand by the machine. Thus the tool stores have to carry stocks of tools in excess of those actually working on the machines.

Regrinding of the tools is done using diamond wheels which are expensive.

The major advantage of using brazed tools is that the cutting geometry best suited for the machining operation can be ground into the hardmetal cutting tip.

Almost all metal machining in the UK was carried out by brazed tools until indexable insert tooling began to be accepted in the late 1950s.

3.2 INDEXABLE INSERTS

In the late 1940s in the USA toolholders were being manufactured with pockets at the nose of the tool in which hardmetal pieces were clamped. The pieces were square, triangular or round. Each corner of the squares and triangles was used for cutting and as one corner became worn the piece was indexed to an unused corner. In the case of a round it was rotated to a clean part of the circumference. These pieces were known as 'Throwaway Tips' which was an unfortunate choice as they still had a scrap value when exhausted. Since the late 1960s they have been known as 'Indexable Inserts'. A selection of simple indexable inserts is illustrated in Figure 15.

Indexable Inserts are now standardised and are designated according to an ISO Standard No. 1832 (1977 and updated in 1985) entitled *Designation of Indexable Inserts for Cutting Tools.*

Under this ISO designation system an insert is described by a cipher made up of letters and numbers. Each letter and number relates to a specific feature concerning the insert and it is possible to have up to 10 features described.

Detail 1. (letter) – Shape of the Insert

The plan view, or shape, of the insert is designated by a single letter.

e.g. square = S
triangle = T
round = R

A parallelogram and Rhombus can have different angles e.g. 55°, 35°, 80° etc. Each of these is described using its own allocated letter.

There are 16 letters in use at the present day.

Fig. 15 Simple Indexable Inserts

Detail 2. (letter) – Clearance Angle on the Insert

The so called 'Clearance Angle' of the insert is the angle which the side of the insert makes with a line perpendicular to the face of the insert. An insert which forms a right angle at the corner where the side meets the face has a clearance angle of 0° and is designated with the letter N. An angle of 11° is described by using the letter P.

At the present day 10 letters are used. 9 of these relate to specific angles in a range from 0° to 30°. The tenth letter (O) is used for clearance angles which do not fall within this range and one would have to refer to the supplier's catalogue for interpretation.

Detail 3. (letter) – Tolerance

A letter is used to describe the accuracy, or 'tolerance' to which the insert is manufactured. It identifies the tolerance on thickness and on the inscribed circle dimension.

By definition one of the main features of indexable inserts is that when they are worn they can be turned to a new position and cutting can recommence because the insert has indexed to the same position it held previously. For roughing and semi roughing operations it is probable that the indexing need not be closer than 0.1 to 0.2 mm. However, for precision turning or for milling, the inserts need to be more accurate and must index so that the cutting edge is repositioned exactly each time.

All indexable inserts are made by a powder processing route and with good pressing technology it is possible to perform roughing and semi roughing operations with inserts which do not have to be ground on the periphery to a smaller tolerance size. When closer accuracy of indexing is required the inserts must be ground and in this case there are bands of accuracy available to meet the needs of the machining operation. The closer the tolerance demanded then the more costly the grinding operation and the more expensive the insert.

There are 12 letters used to designate the tolerance. Three of these letters, U, M and N, relate to inserts unground on the periphery. U has a wider tolerance spread and M and N have a narrower one.

Detail 4. (letter) – Insert Type

There are several ways of clamping inserts into the toolholder or milling cutter body. The simplest inserts are solid and held by an overhead or finger type of clamp, others have a plain hole through the centre of the insert and are pulled back into the pocket of the toolholder by a lever. A further method is to use a screw which passes through the centre hole and locates against a countersink formed in the insert. Some inserts can be turned over and used on both sides and so the countersink is formed into both top and bottom faces of the insert. All these clamping systems are described in Chapter 5.1.9.

Other features which can be incorporated include chipgrooves which are formed into the faces of the inserts.

Thus an insert may be plain (no hole, no chipgroove), this is designated N. An insert which is plain with a chipgroove on one face is designated R and an insert with a plain cylindrical hole and a chipgroove on one face is designated M, etc.

15 letters are used, 14 of which cover specific cases of detail. The remaining one, which is X, is reserved for special types peculiar to any one supplier.

Detail 5. (two digit number) – Insert Size

Because indexable inserts originated in the USA they are dimensioned in inches, e.g. ½″ square, ⅜″ triangle (⅜″ being the diameter of the inscribed circle).

When the ISO standard was drawn up it was decided to quote the dimensions in whole millimetres.

The size of an insert is quoted by reference to the length of the cutting edge when looking at the plan view except for rounds where the reference dimension is the diameter.

A half inch square has a cutting edge length of ½″ (12.7 mm.) and is designated 12 (note that the values chosen are not necessarily taken to the nearest whole millimetre).

Detail 6. (two digit number) – Insert Thickness

As with insert size the thickness is dimensioned in whole fractions of an inch but is designated in millimetres. 9 thickness positions are listed in the standard 01, 02, 03, T3, 04, 06, 07, 09 and 12. They range from 1.59 mm to 12.7 mm thick. T3 is an exception in that a letter is introduced. It was added as an extra thickness position at a later stage.

In order to ensure that the inserts will seat correctly in the toolholder pocket they are either lapped or ground flat. This is the case even with the wider toleranced inserts.

Detail 7. (numbers and letters) – Corner Configuration

In the case of turning inserts the corner configuration is invariably a radius. 10 radii are nominated (with round inserts the symbol adopted is 00). With milling inserts a radius may be used but the more popular situation is to employ a cutting facet ground at the corner. The cutting facets are prescribed in the standard and are designated by letters.

Detail 8. (letter) – Cutting Edge Condition

The treatment of the cutting edge of an indexable insert can have a big influence on its performance. For example a negative chamfer or land, designated T, is of benefit to the toughness of the cutting edge but does not perform well in finishing operations where a sharp edge, designated F, is an advantage.

Up to 6 edge conditions are listed in the standard and each one is designated by a letter.

Detail 9. (letter) – Cutting Direction

The insert may be 'handed', i.e. it can only cut in one direction, or it may be able to cut in either direction.
Inserts are designated:

R = right hand cutting
L = left hand cutting
N = either direction

Detail 10. (a hyphen followed by one or two symbols) – Style of Chipgroove/Manufacturer's Options

This is an optional detail which is added to give additional information concerning the style of a manufacturer's chipgroove. Letters or numbers are used prefixed by a hyphen.

An example of a popular turning insert is given below with an explanation of each symbol. In this example the special chipgroove designation, –11, is a fictitious one.

Example: SNMG 120408ER-11

S = square
N = 0° clearance angle
M = tolerance on inscribed circle ± 0.08 mm.
G = plain cylindrical hole, chipgrooves on both faces
12 = cutting edge length of 12.7 mm
04 = thickness of 4.76 mm
08 = corner radius of 1.2 mm
E = rounded cutting edge
R = right hand cutting
–11 = manufacturer's own chipgroove design 'style 11'

This ISO designation holds good for indexable inserts made from any cutting material. High speed steel indexable inserts are shown in Figure 1.
There are other indexable inserts which cannot be defined by the ISO designation system, for example parting and grooving inserts (Figure 16) and also threading inserts. Such inserts tend to be held in tool-

Fig. 16 Indexable Insert for Parting Off

holders which have standard overall dimensions but have specially shaped pockets specifically made to hold the non standard inserts.

It is essential when machining metals that the flow of swarf which is produced during the cutting operation is controlled. With cast iron workpieces the swarf is formed as small chips which fall from the work-piece and can be carried away on conveyer systems to a collecting station.

When turning steel the swarf produced will usually be long unbroken spirals unless something is done to break them up. Such snarling type chips are dangerous to the machine operator. They can also foul up the movements of tools in CNC machines, cause damage to the cutting edge of the tool and also to the surface of the component which is being machined. Thus when turning steels (long chipping materials) it is essential to break up the chips into manageable pieces.

The so called 'chip breakers' cause the chips to turn over on themselves and then to break up into small curls shaped like a Figure '6' or '9'. The original chipbreakers were in the form of a step ground into the rake face of brazed turning tools just behind the cutting edge. As the chip contacted the back of the step it was made to turn back and eventually break.

Grinding chipbreakers in hardmetal is expensive as it involves the use of diamond wheels and is time consuming. The grinding action must not be too heavy or cracks will result. The advantage of indexable inserts is that grooves, which control the chip, can be pressed into the rake faces.

Fig. 17 Examples of Chip Control Grooves in Indexable Inserts

Fig. 18 Indexable Inserts tipped with Polycrystalline Diamond

An example of this is illustrated in Figure 17 which shows four shapes of hardmetal inserts with differing configurations of chip control grooves pressed into their rake faces.

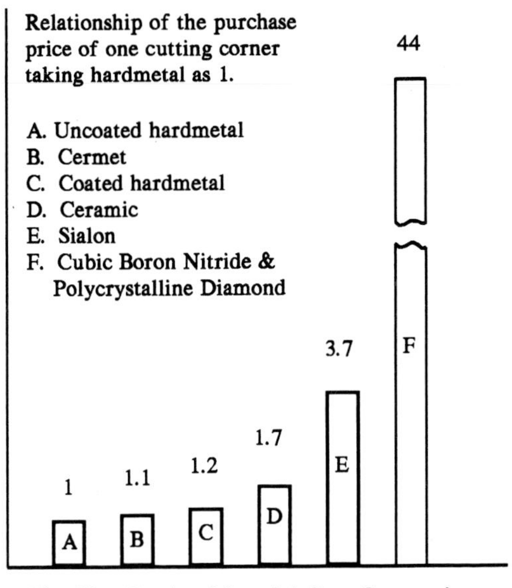

Fig. 19 Cutting Material Cost Comparison

The number of cutting edges available on an indexable insert is governed by its shape and also whether it can be turned over and used on the reverse side. However, with polycrystalline diamond the basic insert is made from hardmetal and one of the cutting corners is tipped with a small piece of the very expensive man made diamond. Examples of PCD tipped indexable inserts are illustrated in Figure 18.

A comparison of the cost of cutting materials can be made by calculating the purchase price of a cutting corner on the same type of indexable insert. A pictorial representation of this is given in Figure 19 where normal uncoated hardmetal is the cheapest per cutting corner whilst polycrystalline diamond is some forty times dearer.

Indexable inserts have brought many advantages in productivity but perhaps the greatest has been in relation to the developments of coatings in hardmetal.

4
Coatings

Coated hardmetals have brought about tremendous increases in productivity since their introduction in 1969. Since that date coatings have also been applied to high speed steel and especially to HSS drills. Coatings are diffusion barriers, they prevent the interaction between the chip formed during machining and the cutting material itself. The compounds which make up the coatings used are extremely hard (> 2500 VDH) and so they are very abrasion resistant. Typical constituents of coatings are Titanium Carbide (TiC), Titanium Nitride (TiN), Titanium Carbonitride (TiCN) and Alumina (Al_2O_3). All these compounds have extremely low solubility in iron and they enable inserts to cut at much higher speeds than is possible with uncoated hardmetals.

4.1 SINGLE LAYER COATINGS

The first coating was a single layer of TiC, 10 to 12 micrometres thick, which was deposited by a process known as Chemical Vapour Deposition (CVD) onto a substrate of hardmetal. During the deposition process some carbon was taken up from the surface of the hardmetal as part of the coating and this changed the carbon balance at the junction of the coating and the hardmetal substrate. This lowering of the carbon balance caused the formation of a brittle compound at the interface between the coating and the substrate and made the early coated indexable inserts sensitive to chipping of the cutting edge.

The next development was to put down a coating of TiN which prevented any decarburising of the hardmetal substrate but the coating, which is gold in colour, did not adhere well to the hardmetal base. TiN is an even better diffusion barrier than TiC but TiC has better abrasion resistance.

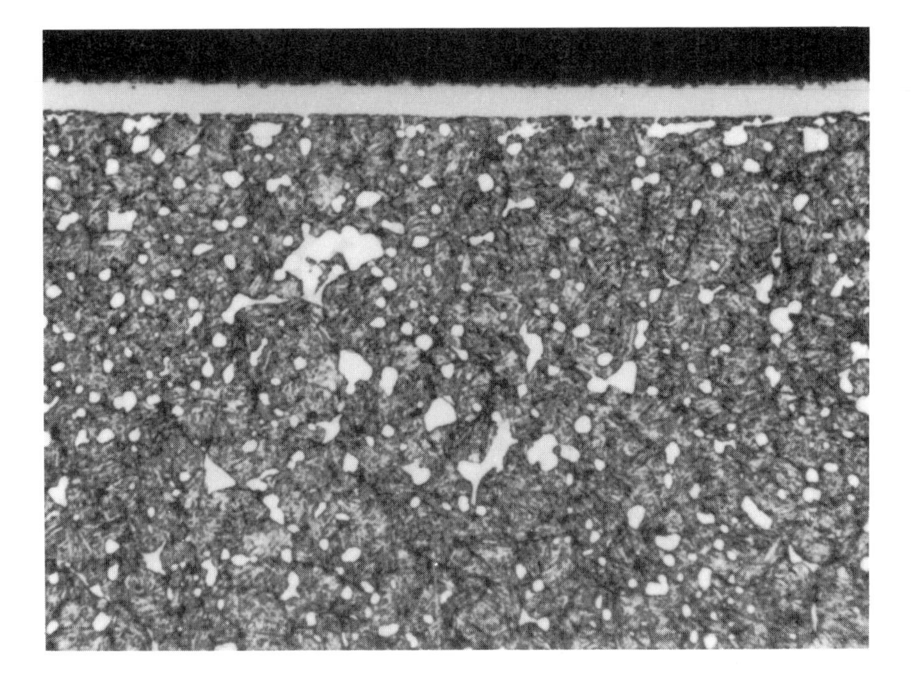

Fig. 20 TiN coating on High Speed Steel × 1000

In contrast to hardmetal, TiN is an excellent single layer coating for high speed steel. Figure 20 shows a cross section of a powder metallurgy high speed steel indexable insert × 1000. The light band at the top of the photomicrograph is the TiN layer which in this case is approximately 5 micrometres thick. The layer is fully dense and is very even in thickness.

The CVD method of coating is carried out at comparatively high temperature ca. 900°C and this will cause high speed steel to soften. Post heat treatment can be carried out but is liable to cause distortion. However, an alternative method exists known as Physical Vapour Deposition (PVD) which is done at a much reduced temperature max. 500°C and allows the high speed steel to retain its hardness. It is a very popular method for coating HSS drills with TiN.

CVD coating gives a much faster rate of deposition than PVD and coatings made by CVD are usually in a thickness range from 5 to 12 micrometres. PVD coatings are much thinner, usually about 3 micrometres, with a maximum of about 5 micrometres.

4.2 TWO LAYER AND TRANSITIONAL LAYER COATINGS

In the early 1970s the problem of adhesion of the coating on hardmetal was resolved by first applying an extremely thin layer of TiC, ca. 0.5 micrometres, which almost completely eliminated the formation of the brittle phase and bonded perfectly to the substrate. The TiC layer then formed a base to build on and a second layer of TiN or of Al_2O_3 could then be deposited. The two layer systems with TiN on the outside perform extremely well on steels and very well on cast irons. The TiC + Al_2O_3 coatings do very well on cast iron because of their higher abrasion resistance but Al_2O_3 is not quite as tough as TiN and so although these coatings do perform quite well on steels their cutting edges tend to be sensitive to the shocks experienced in interrupted cutting.

Some hardmetal producers claimed that with TiC + TiN, stresses existed where the two coating layers were joined. By using the fact that TiC and TiN form a complete series of solid solutions as one moves from pure TiC through the full range of carbonitrides to pure TiN a coating was developed which started with the very thin layer of TiC and then moved through a continuously increasing nitrogen containing carbonitride until it reached pure TiN. The intermediate continuous layer of TiCN between the TiC and the TiN is known as a transitional layer.

It is difficult to produce transitional layer coatings by the PVD method. The CVD process is ideal for transitional layers because it involves gaseous reactions and by altering the compositions and amounts of the gases present it is possible to deposit coatings as desired.

In the CVD process the indexable inserts are placed in a reactor and heated to the required temperature. The reactions take place in a broad range of 750°C to 1050°C. The coating rate is temperature dependent and is much slower in the lower temperature region. Titanium carbide is formed at the surface of the hot indexable inserts by the reaction of hydrogen (H_2), titanium tetrachloride ($TiCl_4$) and methane (CH_4). TiC deposits at the surface and HCl passes on into the exhaust system.

Titanium nitride is formed if the methane is substituted by nitrogen. If a mixture of methane and nitrogen is used then titanium carbonitride is formed. More methane gives a carbonitride of higher carbon content and vice versa.

A photomicrograph of a coating consisting of TiC, TiCN and TiN is shown in Figure 21. The higher carbon content TiCN appears as the dark band, it is purple in colour then changes through pink to orange

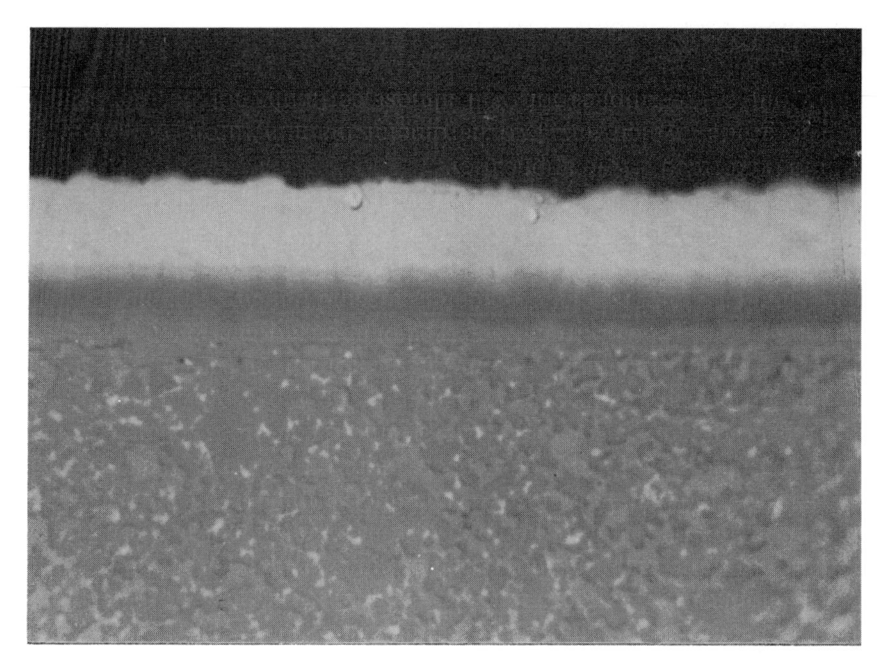

Fig. 21 Tic/TiCN/TiN Coating on Hardmetal × 1500

and finally gold as it approaches the nitrogen rich end of the TiCN transitional layer. The magnification is × 1500 and the coating is about 10 micrometres thick.

A coating of alumina is produced by using hydrogen, aluminium chloride ($AlCl_3$) and carbon dioxide (CO_2) which results in Al_2O_3, HCl and CO.

4.3 MULTI LAYER COATINGS

One of the later developments has been the introduction of multi-layer coatings. These can consist of as many as eight layers within a total thickness of 10 micrometres or less. A typical coating of this type could commence with a thin TiC layer on the hardmetal substrate and then have a TiCN transitional layer moving through to TiN following on with alternate Al_2O_3 and TiN layers ending with an outer surface of Al_2O_3.

A cross section of an indexable insert with such a coating can be seen in Figure 22. The magnification is × 1500 and the thickness of this multi

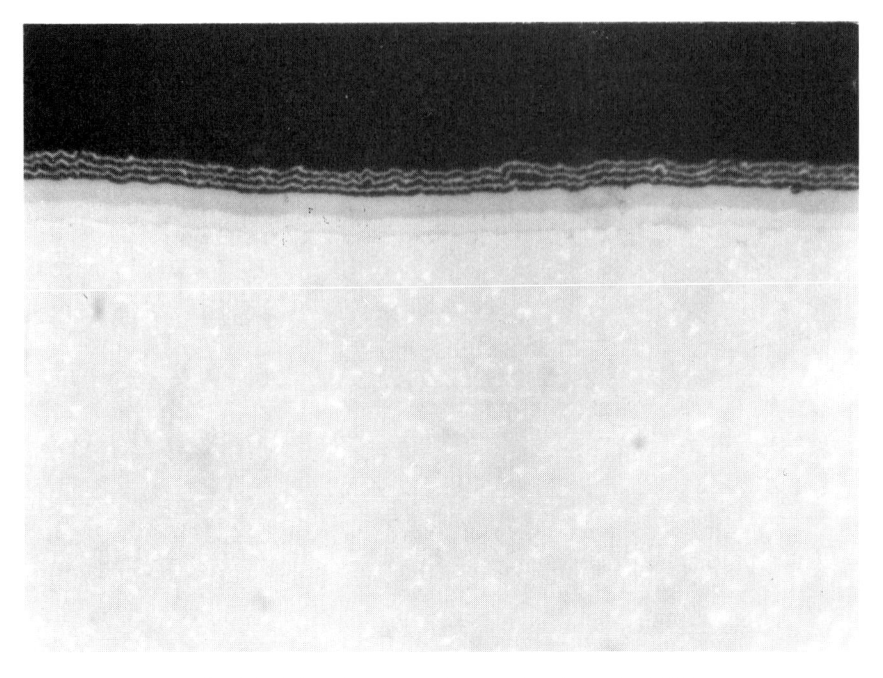

Fig. 22 Multilayer Coating on Hardmetal × 1500

layer is about 8 micrometres. Each of the Al_2O_3 layers is less than 1 micrometre thick and is supported by a TiN layer of similar thickness. This laminated structure is less sensitive to brittle failure than a two layer system of the same total thickness i.e. 4 micrometres TiN + 4 micrometres Al_2O_3. This type of coating has to be deposited by the CVD process. It is unrealistic to think about doing this by PVD.

The early coatings gave improvements in performance over uncoated hardmetal of up to three times. These more sophisticated coatings give improvements up to nine times those of conventional hardmetal.

4.4 COATING STRUCTURE

If the CVD coating process is carried out at the middle and upper end of the temperature band then a columnar structure is achieved and this is illustrated in Figure 23 which is a scanning electron microscope (SEM) photomicrograph X 3000. This type of structure has the disadvantage that the columns can tend to split apart and are therefore sensitive to shock loading. By modifying the coating process, e.g. working at a lower temperature, a granular structure can be produced such as that shown in Figure 24. The magnification is the same as in the columnar structure photomicrograph. This granular structure is advantageous and gives a very reliable performance.

Some manufacturers claim that better edge strength is achieved by first enriching the surface of the hardmetal substrate with cobalt and then depositing the coating on this Co enriched face. This is claimed to make the coating better able to perform in interrupted cutting and to broaden its application field.

4.4.1 COATINGS FOR TURNING

In turning operations CVD coatings are used for both continuous

Fig. 23 Coatings – Columnar Structure × 3000

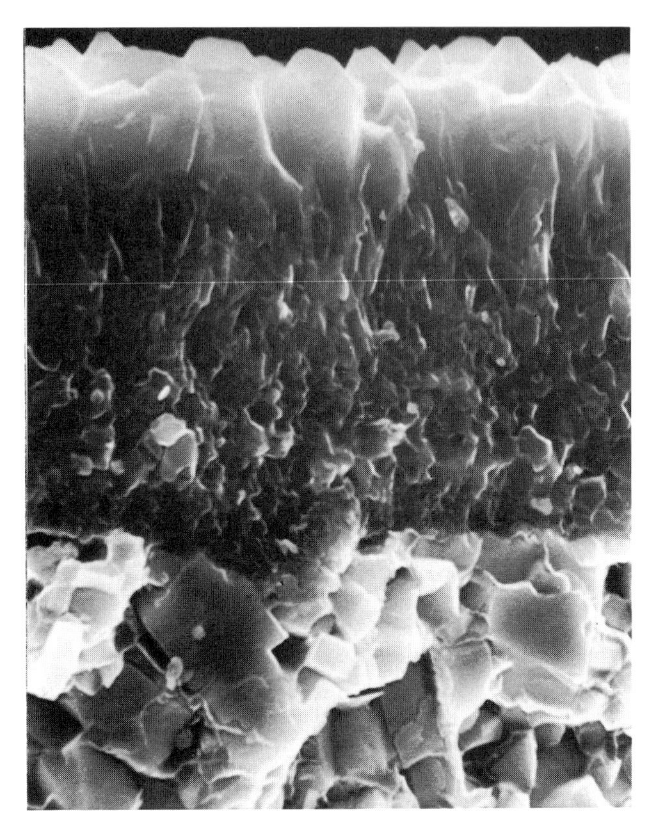

Fig. 24 Coatings – Granular Structure × 3000

cutting and interrupted cutting. In both cases the coating thicknesses supplied by the various manufacturers range from 8 to 12 micrometres.

The hardmetal substrates used are harder for the lighter cuts. For interrupted and heavier cutting tougher hardmetals are used as the base material. Figures 25 and 26 are photomicrographs of the structure of a TiC, TiCN, TiN coating for turning at magnifications of × 1500 and × 4500 respectively. The larger magnification SEM photomicrograph clearly shows the granular structure of the coating.

4.4.2 COATINGS FOR MILLING

Milling operations result in interrupted cutting. The chip thickness is usually quite small and so the shock tends to be concentrated right at

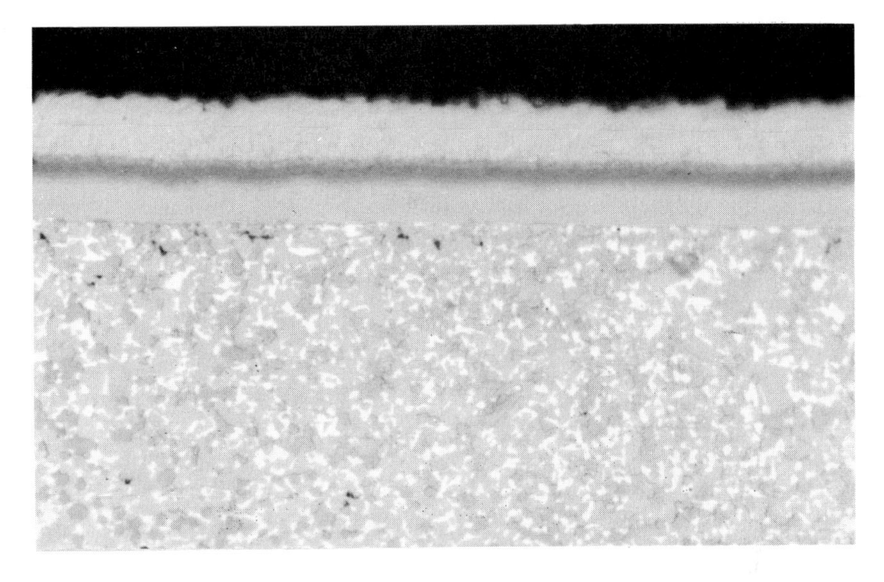

Fig. 25 TiC/TiCN/TiN Coating for Turning × 1500

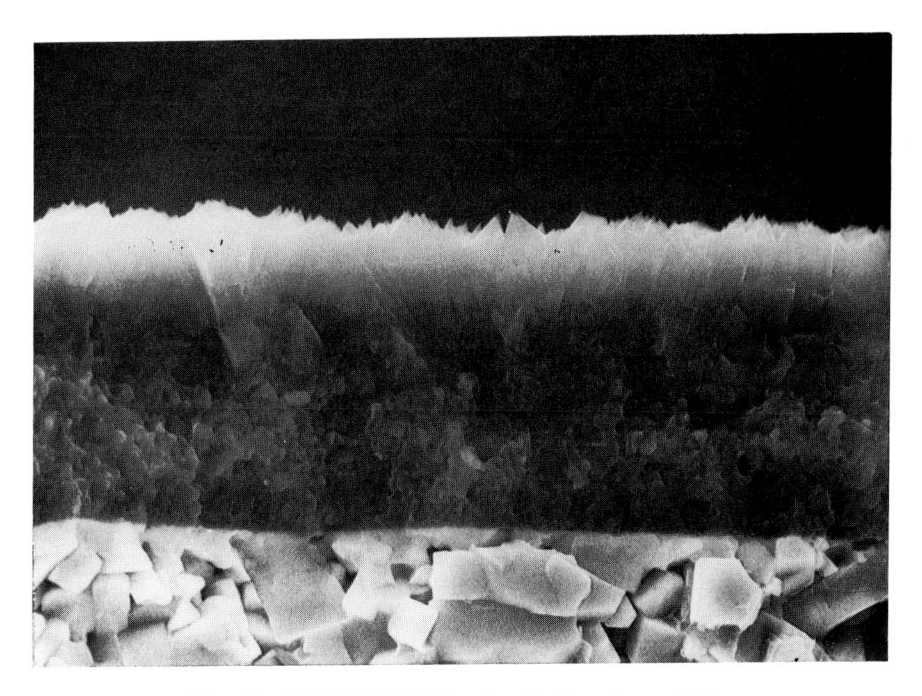

Fig. 26 TiC/TiCN/TiN Coating for Turning × 4500

the cutting edge. A further problem which results from the interruptions is the thermal cycling of the cutting edge as it goes in and out of cut. Thinner coatings cope better with the detrimental effects of milling and thicknesses of 5 to 6 micrometres are typical. The hardmetal substrate must be tough enough to resist the shock from the interrupted cutting and also be designed to combat the thermal cycling. A typical TiC, TiCN, TiN coating is shown in Figure 27 at a magnification of × 1500. An SEM photomicrograph × 4500 of the same coating is presented in Figure 28. Once more the granular structure of the coating is very evident.

4.5 PVD COATING ON HARDMETAL

One of the problems with CVD coatings is the maintaining of the perfectly sharp edge which is sometimes used on hardmetal indexable inserts. This is due to the coating thickness. However, if one attempts to deposit a very thin coating by the CVD process it is difficult to produce a uniform layer and it is also not possible to deposit the sophisticated coatings within the thickness parameters required.

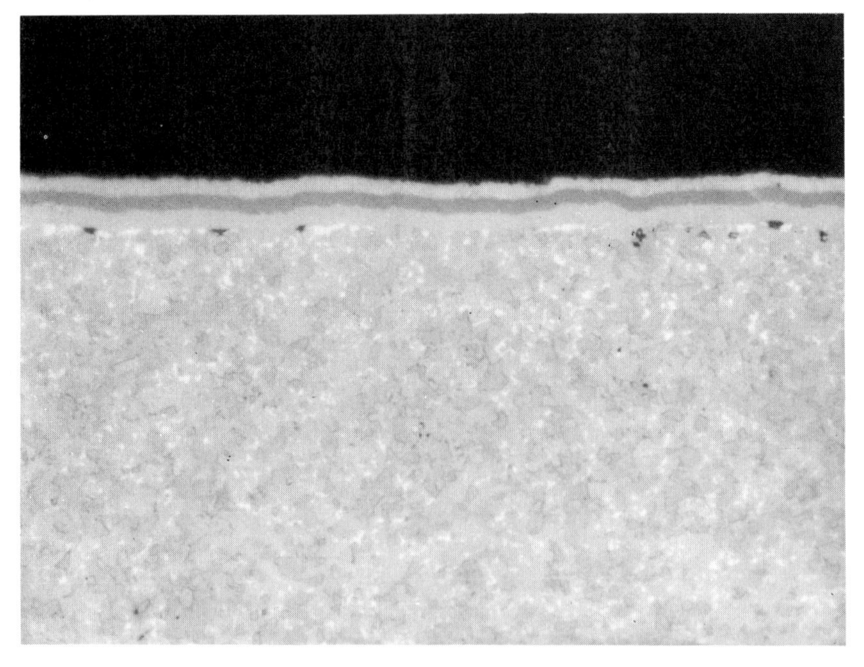

Fig. 27 TiC/TiCN/TiN Coating for Milling × 1500

Fig. 28 TiC/TiCN/TiN Coating for Milling × 4500

Machining operations where only a very thin chip is removed need to be performed with a sharp edged cutting insert. Thread machining is a typical example of this.

Indexable inserts with sharply ground cutting edges are best coated by the PVD process. The coating is then thin enough to maintain a keen edge at the cutting corner. Figure 29 shows a titanium aluminium nitride (TiAlN) coating deposited by PVD at a magnification × 4000. The coating is very uniform in thickness, 2 to 3 micrometres.

4.6 HOW DO COATINGS WORK?

Coatings are not just hard layers which can be applied to any material and improve its abrasion resistance so that it will perform as a cutting tool. Their action in cutting is more sophisticated than that.

They Reduce Cutting Forces

When machining steels under controlled conditions measurements using a toolpost dynamometer clearly show that identical indexable

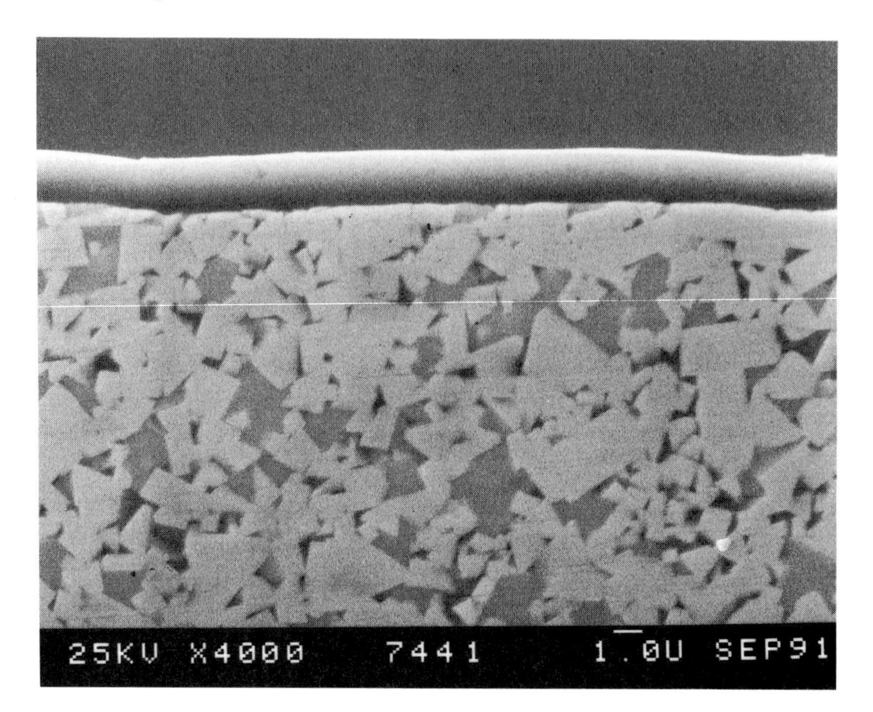

Fig. 29 TiAlN Coating on Hardmetal by PVD × 4000

inserts, except for coating, give higher cutting forces when machining with an uncoated insert. The recorded traces show lower forces and much less fluctuation of these forces when using coated inserts.

They Reduce the Cutting Edge Temperature

Thermocouples inserted into holes spark machined to within 0.25 mm. of the cutting edge of identical uncoated and coated inserts recorded 1000°C in the case of the uncoated insert and 800°C with the coated insert. The steel workpiece and the machining parameters were the same in each case.

They Increase Abrasion Resistance

The coatings are considerably harder than normal hardmetal and so the cutting edge has more abrasion resistance but this is only one of several factors which go to make coated indexable inserts successful. The other advantage of their abrasion resistance is that they do not rub away as the chip is flowing over the rake face.

They Are Diffusion Barriers

The materials which are used as coatings all have an extremely low solubility in iron at the temperatures which arise during machining and so no cratering occurs as a result. When a steel chip is moving over the rake face of a coated indexable insert the coating acts in the same way as a lubricant. There is no linking of the chip to the insert and so friction is much reduced. This in turn means that temperatures are less than with uncoated inserts.

4.7 WHY DON'T COATINGS BREAK?

As already stated the materials which form the coatings are extremely hard, TiN is > 2600 VDH, TiC and Al_2O_3 are > 2800 VDH. With a coating, unlike hardmetal, there is no softer, tougher binder to 'cement' these hard materials. The secret behind their ability to stand up to the cutting loads is a combination of the thickness of the coating and the compressive strength of the substrate on which it is deposited.

No one would regard glass as being other than a very brittle, non-ductile material but if we take an extremely thin sheet of glass, say 2 mm thick × 800 mm long, and support it at each end and then apply a load in the centre the glass will take a considerable deflection. As long as the deflection is not too great then the glass will not break and will restore to its original position when the load is removed.

If we use thin coatings, < 15 micrometres, they will deflect under load but they must not be allowed to deflect too far. If we support them with hard substrates which can themselves take the loads without deforming or breaking then the coatings will remain intact.

4.8 DESIGN OF COATINGS

The early stages of design concentrated on the problems of ensuring that the coating was well adhered to the hardmetal substrate. At the same time the performance of TiC, TiN, HfN, TiCN, Al_2O_3 and various combinations of these compounds was evaluated. In addition the structure of the coating itself – columnar or granular – has been thoroughly investigated and the way in which thickness plays its part has been determined. There are so many possibilities which can arise from these investigations and so it is not surprising that each manufacturer has tended to follow his own chosen route to arrive at the coatings he offers.

One of the major difficulties with developing cutting materials is that there is no single machining test which will simulate what really happens in practice. In turning operations the workpiece material, the workpiece shape and its configuration, the cutting speed, the feed, the depth of cut, the cutting geometry, the treatment of the cutting edge, the rigidity of the machine tool and other factors have considerable influence on the way a cutting material will perform. Laboratory tests will certainly assist in proving which materials should do well in the field but the ultimate test is what happens on the shop floor. It takes time to assimilate the real performance of a cutting material.

The early coatings were usually deposited on existing hardmetal indexable inserts on the principle that coating them would improve their performance. However, a coated insert should be regarded as a two part system – the coating and the substrate.

The job of the coating is to combat the metallurgical reactions which take place at the chip/tip interface and also to combat abrasive wear.

The substrate has to cope with the mechanical aspects involved in the machining operation. It must withstand the basic cutting load and any shock involved so that chipping and breakage do not occur. It must also have high hot strength to resist plastic deformation of the cutting nose as the temperature rises during cutting. Because the coatings themselves are crater resistant it is not necessary to add large amounts of TiC to the hardmetal substrate and this is a big advantage from a toughness point of view. It should be appreciated that the hardmetal substrate never 'sees' the chip and so some of the things that are done in the case of uncoated inserts can be modified when coating is involved.

Thus the design technology involved for any particular type of machining should be to optimise the coating and then to optimise the substrate. For example:

When milling an automobile crankshaft, a standard coating 6 micrometres thick deposited on a standard substrate produced 60 components before the inserts had to be indexed. It was possible to optimise the substrate by giving it better hot strength and higher toughness. Using this new substrate but keeping the original coating 114 components were produced for each indexing. The coating could then be made more resistant to the repeated shock of the milling operation by reducing its thickness to 3–4 micrometres. When this was done the combination of the two improvements resulted in 215 components being produced for each indexing of the milling inserts.

Summary

Coated hardmetal indexable inserts give remarkable improvements in performance over uncoated inserts. When used at the same cutting speeds normally employed for uncoated inserts they will reduce the frequency of tool indexing and give savings in down time and tool costs.

However, their real advantage is that they will cut at faster speeds than uncoated hardmetal inserts and therefore produce considerably more components in a given time. With the expensive machine tools in use today this has a big effect in reducing production costs.

Another advantage of coated hardmetal inserts is that each grade has a broader application range than the equivalent uncoated insert. This can reduce the need to stock a wide range of grades.

The coatings and substrates can be optimised to cope with the problems arising in any machining task. Each manufacturer has developed his own coatings and substrates to produce the properties required from an insert. There are no standard compositions of coating or of substrate.

Coating of brazed hardmetal tools is not a practical solution. At CVD coating temperatures the usual brazing solder will melt. An even more important point is that even if copper brazing is used then regrinding of the cutting edge and chipbreaker will remove the coating and recoating between regrinds is out of the question.

Most machining operations are now carried out using tooling systems which have indexable inserts as their cutting tips. It is estimated that 80% of these inserts are coated.

5
Machining Processes

5.1 TURNING

Turning is probably the most used of all the machining processes. About one third of the machines in production are employed in turning. The continued developments in cutting materials and in cutting tooling keep turning to the fore as an economic method of manufacture.

This section of the book will deal with turning, both external and internal (boring). It will include comments on cutting parameters, chipgrooves, toolholders, insert clamping systems, workpiece materials and failure mechanisms.

Parting off, grooving and threading which are usually regarded as part of the family of turning operations are covered in separate sections.

5.1.1 CUTTING PARAMETERS

Only those features which have a major influence on the way a tool performs when turning will be dealt with.

Figures 30 and 31 define some of these features and should be referred to when reading this section of the book. The subject used in both diagrams is a representation of a toolholder fitted with a square shaped indexable insert which is turning a round bar.

a) Cutting Speed (v_c)

The most important parameter is the cutting speed. This can be defined as the speed at which the workpiece is passing over the cutting edge. It is normally quoted in metres per minute (m/min). The cutting speed (v_c) is calculated from the following formula:

$$v_c = \frac{d \times \pi \times n}{1000}$$

where d = the diameter of the workpiece at the point of cut in mm
n = the number of revolutions per minute of the workpiece
v_c = the cutting speed in metres per minute

The workpiece material which is being machined has a direct influence on the selection of the cutting speed to be used. For example it is possible to cut soft materials at high speeds on most of the machines used for turning. However, when the workpiece material is extremely hard there can be a limitation on cutting speed because the cutting forces generated at high speeds will need more power than the machine can deliver, the driving mechanisms may become overloaded and the holding systems could fail.

The possible cutting materials which can be used to perform the turning operation can also influence the choice of cutting speed. For example, high speed steels can only work at lower speeds, up to 50 m min^{-1}, these are followed by uncoated hardmetals which for general turning operate around an upper limit of 150 m min^{-1}. Coated hardmetals can work at higher speeds of the order of 200 m min^{-1} and cermets can perform at 400 m min^{-1}. Ceramics cut at around 450 m min^{-1} in their upper speed range whereas cubic boron nitride performs at 600 m min^{-1} and finally, polycrystalline diamond cuts at speeds of 1200 m min^{-1}. It must be remembered, however, that these cutting materials can only perform at the speeds quoted on specific workpiece materials. In general, for any given workpiece material, as the cutting speed increases so the cutting force also increases and with it the temperature of the cutting edge.

b) Feed (f)

The feed is the relative movement of the tool in the direction of the workpiece axis and is expressed as the distance moved in one revolution of the workpiece. It is quoted in millimetres per revolution (mm/rev).

As the feed increases the chip cross section increases and so the cutting force is also increased. This in turn increases the stresses imposed on the machine and on the workpiece. The strengths of the machine, the workpiece and the holding device may therefore limit the allowable feed.

Feeds are often termed as 'coarse' or ' medium' or 'fine' and this has a relevance to the finish which they produce on the workpiece. A large, or coarse feed will tend to leave a finish like a screw thread on the workpiece whereas a small, or fine feed will leave a much smoother finish.

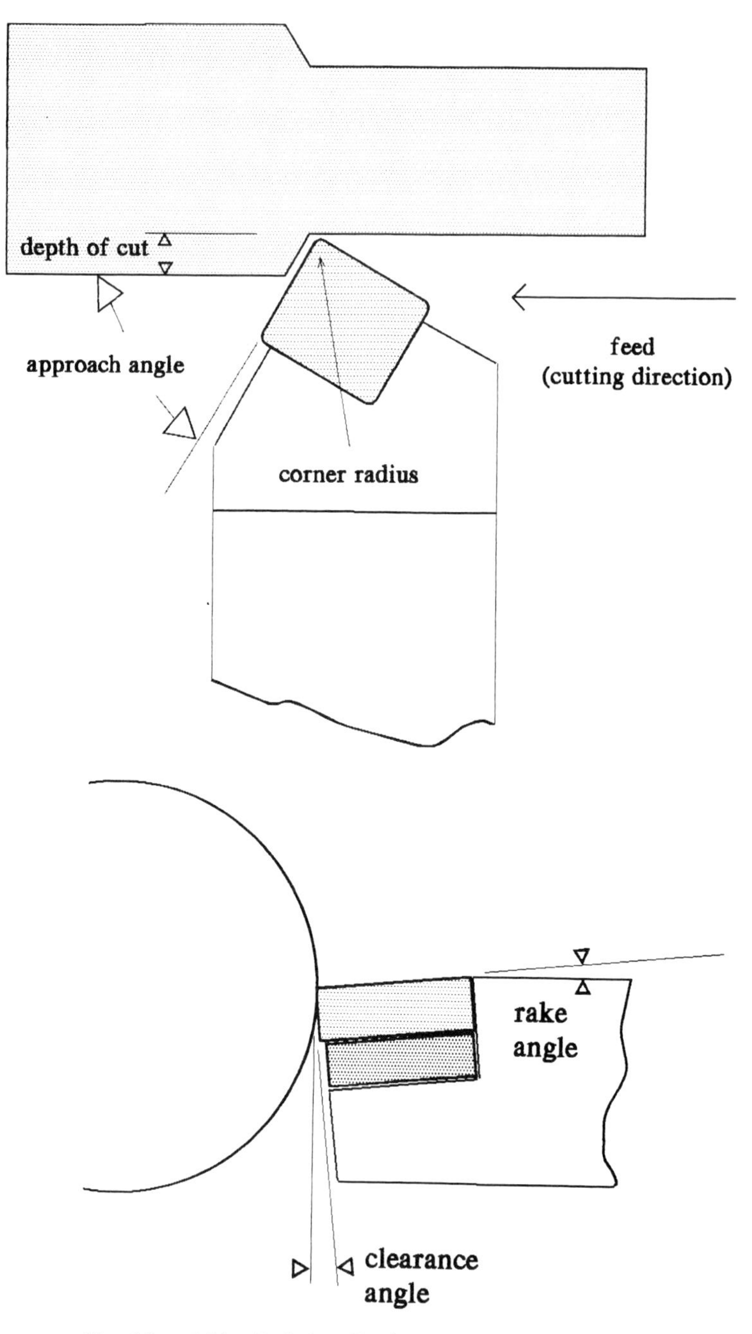

Figs 30 and 31 Defining Cutting Terms in general use

c) Approach Angle
The approach angle is illustrated in Figure 30. In any turning operation this angle should be established first. An approach angle of 75° is satisfactory for most standard turning operations. If the machining operation is the production of a square shoulder then an approach angle of 90° is needed. This reduces the cutting force in the direction of the tool axis (the 'push off' or 'back' force) and this in turn reduces the tendency to deflect the workpiece which is better for slender components. It is also suitable for finishing operations.

Workpiece surfaces which are very uneven, e.g. a rough forging, are better machined by reducing the approach angle to 70° or even 45°. This protects the cutting edge at the start of the cut and helps when performing roughing operations.

An approach angle of 30° is better for machining hard materials. It produces small chip widths and thus lower specific cutting edge loads which in turn reduce wear. The disadvantage of this low approach angle is that a high push off force results. This makes it necessary to have good stability of the workpiece, the machine and the holding mechanism.

Approach angles greater than 90° are used when turning and facing with the same tool as in copy turning and producing undercuts. In this case care must be taken to avoid breakage of the nose of the tool.

d) Depth Of Cut (a_p)
The depth of cut (a_p) is the distance the tool penetrates radially into the workpiece when performing the turning operation. In finishing operations less metal is to be removed and so the depth of cut is small. With roughing operations, particularly on castings or heavy forgings, the depth of cut is much larger. The depth of cut is not the width of the undeformed chip. Only when the approach angle is 90° does the width of the chip equal the depth of cut, at all other times it is greater.

The undeformed chip cross section is determined by the depth of cut (a_p) and the feed rate (f).

e) Nose Radius (Corner Radius)
The nose is the weakest part of any tool and should be protected against shock, see Figure 30. To protect the nose, try to use an approach angle which is sufficiently small to ease the tool into the work. This is especially important when making interrupted cuts.

Too large a radius tends to cause chatter. A large radius applied to a non-rigid workpiece will chatter due to the substantial wedging action between the nose radius and the workpiece. However, a relatively large radius is permissible if both the tool shank and the workpiece are rigid. A large corner radius improves the stability of indexable inserts and is recommended for roughing applications. Large corner radii can achieve better surface finishes compared with smaller radii when operating at the same feed rate. However, in the case of indexable inserts with chipgrooves large radii are not used for finishing operations because of the chipgroove/corner radius configuration (large corner radii have wide chip grooves which are unsuitable for the smaller feeds used for producing good finishes).

Unstable parts with a tendency to vibration should be machined with a smaller nose radius. Radii which are too small should be avoided. A very small nose radius, less than 0.125 mm, is liable to cause breakage.

f) End Cutting Angle

The end cutting angle is the angle made between the axis of the workpiece and the trailing edge of the cutting tool. This angle should be large enough to allow the tool to clear the work. It should never be larger than necessary because the larger the angle then the weaker is the nose of the tool.

In the case of indexable inserts this angle is determined by the choice of the insert shape and the approach angle of the toolholder being used but with brazed tools the end cutting angle can be varied at will.

An end cutting angle of between 8° and 15° is satisfactory on most standard turning and facing operations. Increasing the end cutting angle tends to reduce chatter because it reduces the pressure on the workpiece. When the workpiece is not very rigid and chatter is encountered the end cutting angle may be increased to 20°.

g) Rake Angle

The rake angle is illustrated in Figure 31. With indexable insert tooling rake angles depend on the choice of insert which can be either positive, neutral or negative.

Positive rakes give low cutting forces and reduce vibration. Swarf flow is easier with positive rake angles. Their disadvantage is that they result in a weaker cutting edge and increase the danger of breakage.

The advantages of negative rake angles are that they produce stronger cutting edges and are therefore suitable for interrupted cutting. They also allow double sided indexable inserts to be employed which in turn means twice as many cutting edges available per insert compared with positive or neutral rake inserts.

On the other hand they give rise to higher cutting forces and have a higher power requirement.

The rake angle is dependent on the application e.g. interrupted cutting etc. and also on the workpiece material being machined. It is also dependent on the cutting material, for example ceramics will not perform with positive rake angles and high speed steel is totally unsuitable with negative rake angles.

h) Clearance Angle

The clearance angle is defined in Figure 31. A clearance angle of 6° to 7° is large enough to prevent excessive rubbing of the tool on the workpiece. Clearance angles that are too small cause rubbing and prevent free cutting especially when very coarse feeds are employed. Clearance angles that are too large make the tool weaker and encourage chipping. They also tend to cause the tool to chatter.

With lower strength workpiece materials such as wood, plastics, non-ferrous metals etc. the clearance angle can be increased.

When machining very hard workpiece materials clearance angles of 4° are best.

If internal machining (boring) is being carried out higher clearance angles are recommended.

i) Wedge Angle

The wedge angle is the angle formed by the rake face and the clearance face. If the rake angle is added to the clearance angle and the total is subtracted from 90° then this gives the wedge angle. The workpiece material to be machined influences the angle of the wedge. Materials which are harder and have a high tensile strength need larger wedge angles. Soft materials are best machined with a smaller wedge angle.

Comment

Because some tools are ground and others are fitted with indexable inserts the remarks about angles given above must be read in relation to the type of tooling involved. In the case of high speed steel, stellite, hardmetal brazed tools and even some special diamond tooling all the cutting angles can be ground to suit the requirements of the turning operation to be performed.

When indexable inserts are used in the tools many of the angles are controlled by the shapes of the inserts themselves.

5.1.2 CUTTING 'MECHANISMS'

The starting point for all considerations of metal machining is the workpiece material. The properties of the workpiece material influence the choice of cutting material, of cutting geometry and of cutting parameters.

For example, if we consider butter as a workpiece material and we take a steel butter knife as the cutting tool then the properties of the butter decide on the way we will cut it. If the butter is soft then we press the cutting edge of the knife down into the butter and little if any resistance to the cutting action is felt and the power required to do the cutting is extremely low. If the butter is hard, having been in the refrigerator for several hours, and we attempt to do the identical operation then much more pressure is needed and the sharpness of the knife becomes relevant. If we want to remove some butter we push the knife sideways with a negative rake action and shear a chip of butter from the hard block. The sheared chip is similar to that produced when machining steels. This negative rake cutting requires a little power and if we take an increased depth of cut to have more butter then the power needed is greater. This analogy is a very simple way to introduce the subject of cutting mechanisms.

In any turning operation there are three forces acting on the cutting tool. The first and most important force is that which is produced by causing the chip to shear away from the workpiece, it acts vertically down onto the rake face of the tool at the cutting edge and is known as the 'main cutting force'.

The second force is that which opposes the feed of the tool as it travels along the surface of the workpiece. Its magnitude is governed by the rate of feed and it is known as the 'feed force'.

The third force opposes the radial movement of the tool into the workpiece and is termed the radial or 'push off force'. Figure 32 illustrates these forces.

The sharpness or acuteness of the wedge angle of the cutting tool influences the main cutting force. If we assume that the clearance angle of a turning tool is 6° then a wedge angle of 84° will result in neutral rake machining. As the wedge angle is reduced so the rake angle becomes more positive. As a rule of thumb, for each 1° positive rake

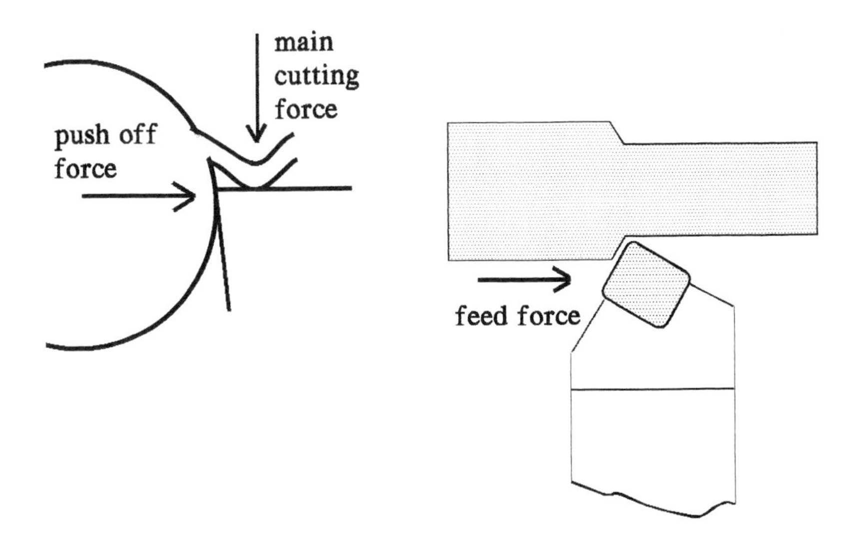

Fig. 32 Cutting Forces – Definition

applied the main cutting force is reduced by 1.5% and so less heat is generated at the cutting edge. However, as the wedge angle is reduced so the strength of the cutting edge is reduced and the danger of breakage of the tool is increased.

Soft Workpiece Materials (Aluminium & its alloys, copper, brass etc.)

When cutting so called soft materials positive rake angles are preferred. They allow the chip to flow freely and they produce much lower cutting forces than those which arise with negative rakes. Sharp cutting edges are also advantageous. If negative rakes are used the swarf tends to ball up on the rake face and as the cutting speed increases this balling up becomes almost unmanageable. A further disadvantage with negative rakes is that the push off force increases and this can cause the workpiece to be deflected or deformed.

Because positive rakes give rise to lower cutting forces cutting speeds can be increased. This increase in cutting speed causes wear of the cutting edge and more abrasion resistance is required from the cutting material. High speed steels quickly lose the sharpness of their cutting edges at these higher speeds and this would also apply if stellite was used. The tougher, higher cobalt grades of hardmetal do not have enough abrasion resistance for the higher speeds and so the low cobalt content, fine grain tungsten carbide grades having a hardness of 1600 VDH and above are selected to machine this group of workpiece materials.

With soft workpiece materials cermets, ceramics and CBN offer no cutting advantages over the harder grades of hardmetal and in any case they are basically not offered with a wedge angle less than 90°.

When extremely high cutting speeds can be employed and very high surface finishes are important, especially on aluminium alloys, then PCD is the cutting material to use. The unique hardness of diamond provides the wear resistance necessary to perform at these very high cutting speeds.

Another point which is important concerning positive rake cutting angles is that because the main cutting force is reduced less power is needed to perform the turning operation and by using reduced feeds machines which have comparatively low power and less rigidity can be employed even on harder workpiece materials. Positive rakes also create lower push off forces and they are therefore ideal for machining components which are long and slender and which tend to bend during cutting if a high push off force is generated. The same comments apply to thin walled workpieces.

Unhardened Steels

The ideal and most efficient way to machine unhardened steels is by using cutting tools with negative rake angles. In this case the requirements are a cutting material with a high hot compressive strength and a machine with adequate power and rigidity coupled with chucking and gripping systems which will hold the workpiece securely during the machining operation.

Effectively the workpiece is driven onto the cutting tool which just stays there as a rigid obstruction. The power transmitted by the machine continues to move the workpiece forward and metal is sheared away from it at the face of the cutting tool. With negative rake the wedge angle of the cutting tool is at least 90° which provides a very strong corner. Additional strength can be given by applying a radius or a chamfer along the cutting edge which has the effect of making the wedge angle more obtuse right at the point of cutting.

Negative rake cutting geometry causes most of the main cutting force to act as a compressive load on the cutting tool. Thus cutting materials

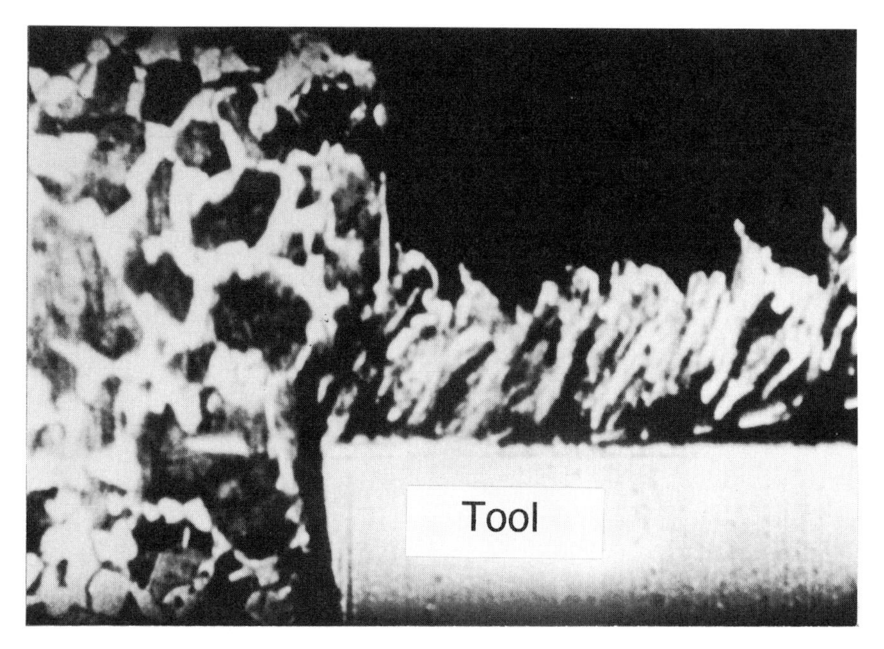

Fig. 33 Chip Formation when Turning Steel

used to perform in this way must have a high compressive strength at room temperature and particularly at the higher temperatures which are generated during cutting. As the cutting speed is increased so the temperature at the cutting edge increases and more power is required to shear metal away from the workpiece. It is at this stage that ceramics begin to perform better than hardmetals because they have higher hot strength.

Negative rake geometry is also necessary if the profile of the workpiece gives rise to interrupted cutting and in this case the tougher hardmetals are essential as cutting materials.

Hardened Steels And Tool Steels

If hardened steels are to be cut with hardmetal then some positive rake and lighter cuts with reduced cutting speed are necessary. However, CBN can be used very successfully with negative rake geometry. In this case the cutting mechanism changes in that the area of the workpiece in contact with the CBN cutting tool becomes so hot that some softening occurs and the chip becomes easier to remove. The extremely high hot strength of CBN enables it to stand up to these very high cutting

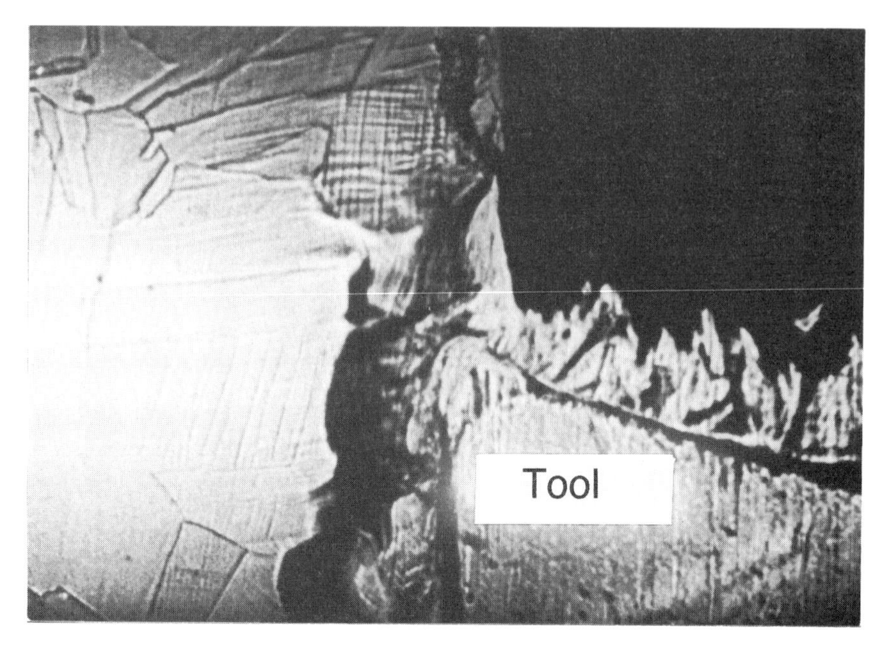

Fig. 34 Formation of Built Up Edge when Turning Steel

temperatures and the negative rake geometry also permits interrupted cutting to be carried out.

Chip Formation

Cross sections showing chip formation are illustrated in Figures 33, 34 and 35.

Figure 33 shows a macro photograph of a hardmetal indexable insert machining a medium carbon steel with neutral rake. This is a still shot of an actual cutting operation.

Figure 34 is a similar macro photograph of turning with positive rake but showing a situation which can arise when machining softer, 'stickier' workpiece materials. This is known as 'built up edge'. Built up edge comes about when a minute seizure occurs between the chip and the hardmetal cutting material right at the cutting edge. Further welding on of the workpiece material then takes place and builds up on the cutting edge.

Built up edge is dangerous, it can suddenly break away taking some of the cutting edge with it and presenting the consequent possibility of breakage of the cutting tool. It also causes a poor surface finish on the workpiece.

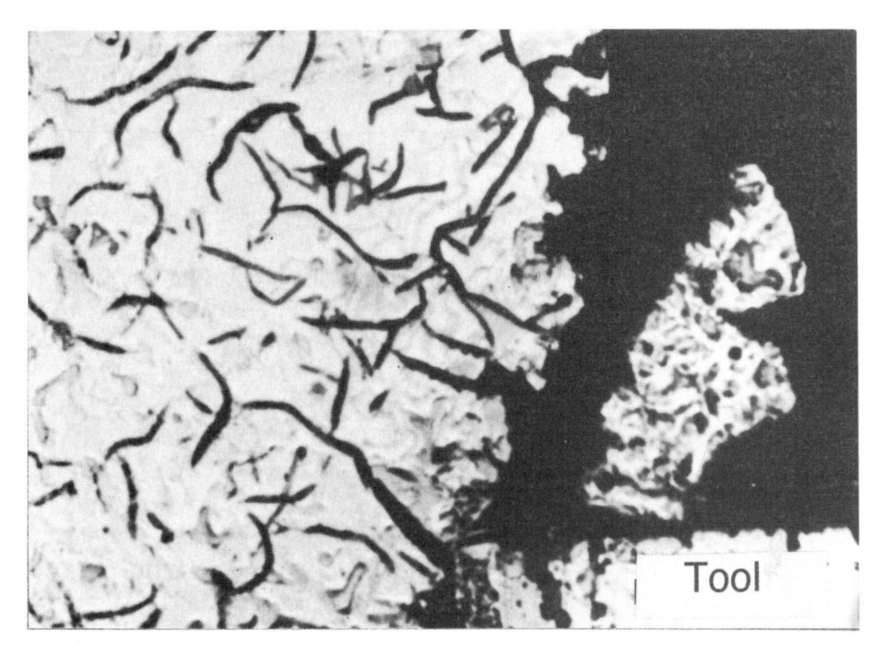

Fig. 35 Chip Formation when Turning Cast Iron

The materials which tend to form a built up edge are low carbon and free machining steels, stainless steels, high temperature alloys, aluminium and titanium. Negative rake geometry and low cutting speeds are the usual promoting factors for built up edge.

The most popular solution to the problem is to increase the cutting speed. Other changes can be the use of a coated grade of hardmetal or to employ a positive cutting geometry. On very light machining a cermet might be used. With titanium it is important to maintain a sharp cutting edge. As a final resort the cutting speed should be increased drastically and copious quantities of coolant should be applied.

The third macro photograph, Figure 35, shows a simple cast iron workpiece being machined by a hardmetal indexable insert. One can clearly see how the chip is being dislocated along the lines of the black graphite flakes present in the cast iron and why normal cast iron is termed a short chipping material and needs no chip control groove behind the cutting edge.

A considerable amount of debris is generated when machining cast iron and much of this passes down the clearance face of the cutting tool causing high flank wear of the cutting material. As this flank wear increases the area of the clearance face in contact with the workpiece increases and the extra rubbing action which results causes a rise in temperature. Those cutting materials which have very high abrasion resistance and high hot strength will therefore perform best on cast iron and it is in this field that ceramics are outstanding.

5.1.3 CHIP CONTROL

When turning long chipping workpiece materials the swarf which is produced can cause severe problems unless it is controlled. The first and most important consideration concerning the swarf is the safety of the machine operator. In workpiece or cutting tool handling situations snarling chips are very dangerous and every attempt should be made to prevent them from being formed. They can also severely hinder automatic functions such as gauging, loading, unloading and tool changing. Chips which turn onto the workpiece can damage its surface finish and can also foul up onto the cutting edge with the possibility of chipping or breakage of the cutting material.

Before the advent of indexable inserts the swarf was controlled by grinding a chipbreaker immediately behind the cutting edges of brazed hardmetal tools. The chipbreaker interrupts the flow of the chip causing

Fig. 36 Acceptable Chips when Turning

it to turn over on itself and break. Ideally the chips should be uniform in size and shaped like a Figure '6'. Such chips fall away easily into the bed of the machine and can be removed by conveyors. An illustration of acceptable chips is given in Figure 36.

Early indexable inserts had no chip control grooves pressed into the rake face and relied on a loose chipbreaker which was part of the tool-holder assembly and which was clamped onto the top face of the insert with its leading edge set at a specific distance from the cutting edge.

Since that time, chip grooves have been developed which are directly pressed into the indexable inserts. The latest chip grooves are computer designed and will perform satisfactorily over a specific range of feeds and depths of cut. Each manufacturer has followed his own experience in perfecting his chipgroove designs and so there are no standard profiles in existence. Figures 37, 38 and 39 show three chipgroove designs from one manufacturer which cover the full range of plain turning operations and serve as an example of the sophistication which has been reached in this field.

The indexable insert illustrated in Figure 37 is intended for finishing operations. The feeds and depths of cut which are used for finishing are small and the chip flow takes place in the area of the cutting corner, hence the concentration of profile design at the corner.

For medium turning, Figure 38, the chipgroove has to cope with a wider range of depths of cut and the chipgroove design closes in

Fig. 37 Chip Control Groove for Finishing Cuts

Fig. 38 Chip Control Groove for General Purpose Cutting

Fig. 39 Chip Control Groove for Heavy Cutting

towards the corner to deflect the chips from the smaller depths of cut and opens out along the cutting edge to cope with the increasing depths of cut and feeds.

Heavier roughing cuts need a chipgroove to cope with larger feeds and depths of cut. Such a chipgroove is illustrated in Figure 39.

The inclination angle of the insert in the tool determines the contact point of the chip on the rake face and is largely responsible for the direction of the chip flow. Negative inclination angles direct the chip back onto the surface of the workpiece. In order to minimise damage to the machined part, especially with internal or fine machining, positive inclination angles are recommended.

The basic features of a chip control groove are shown in Figure 40.

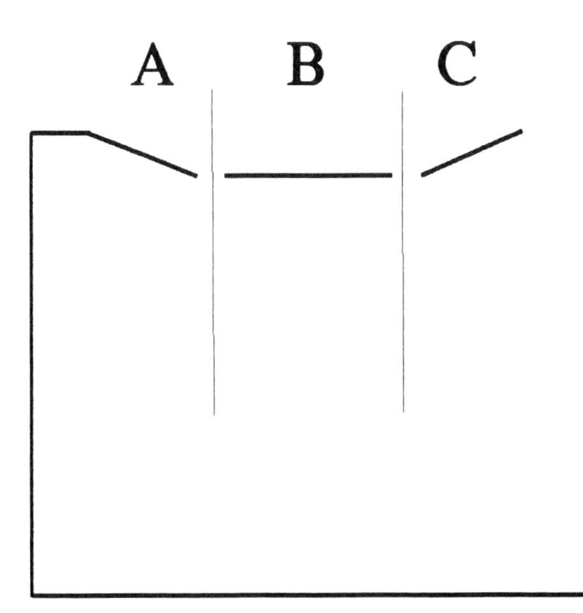

A - Influences Major Cutting Force. For each 1 degree +ve major cutting force is reduced by 1.5%

B - Influences form of chip

C - Influences back force

Fig. 40 Influencing Factors in a Chip Control Groove

Swarf Difficulties and Possible Remedies

Ribbon type chips, stringy and light silver in colour
Increase the feed
Change to an insert with a more suitable chipgroove

Other factors to consider are:

Cutting speed
Depth of cut

Tightly curled, dark blue heavily burnt chips
Reduce the feed
Change to an insert with a more suitable chipgroove

Other factors to consider are:

Cutting speed
Depth of cut

Surface finish not acceptable
Consider effect of changing cutting speed
Consider effect of changing feed
Try alternative nose radius

Other factors to consider are:

Coolant
Centre height of cutting tool

5.1.4 CUTTING EDGE CONDITION

In most turning operations a sharp cutting edge is not the best condition to have because of the greater tendency for breakage. However, with softer materials such as aluminium and with plastics, hard rubber etc. a sharp edge is needed. It is also advantageous with titanium alloys which work harden easily. Again it must be emphasised that a sharp edge is susceptible to damage from any shock loading that may occur during cutting.

The cutting edge of an indexable insert can be given additional protection by applying a facet or by edge rounding. There is a slight increase in cutting force if this is done.

After brazed hardmetal tools have been ground to a sharp edge it is preferable to lightly stroke that edge with a diamond, or boron carbide hand lap. This should be just sufficient to take the keenness from the edge and if this is done the life of the tool can be significantly increased.

In the ISO designation system for indexable inserts, ISO 1832, the cutting edge condition is designated by a letter. The four most popular cases are:

A Sharp Edge – ISO 'F'

A sharp edge gives rise to the lowest cutting forces. It therefore results in the lowest cutting temperature when machining any workpiece material. However a sharp edge is prone to chipping and breakage and therefore should be restricted to continuous cutting. It is not a suitable condition for ceramics and CBN. It is the preferred edge for fine finish machining and will give the highest surface finish. It is most suited to machining cast iron, soft yet tough materials, non-ferrous metals, wood, plastics, hard rubber and composite materials.

The sharp edge condition is not applied on ceramics or CBN and is not recommended for cermets.

Edge Rounding – ISO 'E'

Several techniques exist for applying a small radius or for rounding off the cutting edges of indexable inserts. Smaller rounding is done where the machining operation is light and heavier rounding is carried out where roughing operations are to be performed. The radius applied should always be smaller than the feed which will be used in the turning operation and typical edge radius sizes range from 0.02 to 0.08 mm.

A lightly rounded edge is essential on CVD coated hardmetal indexable inserts used for turning. These have thicker coatings than those deposited by PVD. Coated hardmetals are always slightly more sensitive to shock than uncoated ones.

Edge rounding is the popular condition for hardmetals used for turning steels as it gives some protection against interrupted cutting and possible damage from swarf.

A very light edge rounding is usually given to cermets but this edge condition is not applied to ceramics or CBN.

Chamfer/Negative Land – ISO 'T'

When a higher level of protection is required then a chamfer is applied to the cutting edge. Increasing the angle or the width of the chamfer will strengthen the edge of the indexable insert still further but will also increase the cutting force which could have a deleterious effect. The feed should always be greater than the chamfer width except when machining hardened steels and hard cast iron. Probably the most popular chamfer is 0.2 mm wide at an angle of 20°. For very fine finishing using Al_2O_3 ceramics then a chamfer of 0.05 mm at 20° would be suitable.

With hardmetals this chamfered edge condition is used when interrupted cutting and impact occur. It is also the normal edge condition for ceramics and CBN.

Chamfer Plus Edge Rounding – ISO 'S'

By very lightly rounding off the obtuse corners created at each side of the chamfer where the chamfer meets the rake face and where it meets the clearance face of an indexable insert any tendency to chip or flake at these corners is minimised. This edge condition is particularly recommended when rough turning of steel with ceramics.

After a period of service, railway wheels become heavily work hardened and the surface of the 'tyre' becomes uneven. When this stage is reached the wheels are removed and the tyre is turned to bring it back to its original condition. Hardmetal is mostly used to perform this task and this edge condition is the one normally applied to the special indexable inserts which are used.

This edge condition of chamfer coupled with edge rounding is used for the heaviest cuts and for heavy interrupted cutting with varying depths of cut. It provides the highest safety for the indexable insert but it increases the cutting forces and the temperature at the cutting edge. It also increases the possibility of vibration being set up during turning.

5.1.5 ISO APPLICATION GROUPS

ISO 513 is a standard which classifies applications by first taking workpiece materials and nominating them into one of three main machining groups, designated P,M or K, and then subdividing each of these three groups into machining applications which are identified by a number prefixed by the letter of the main group e.g. P10, K20 etc. A colour coding is also used to help identification of the main groups and is used on brazed tools and on packaging.

The machining applications with lower order numbers, 01 and 10, cover the very light finishing operations and the higher order numbers, 40 and 50 relate to heavy, interrupted roughing operations.

The following information is a simplified interpretation of the standard:

P

Colour code BLUE

Workpiece materials which fall into this main group are:

Steel
Cast steel } including ferritic and martensitic
Long chipping malleable cast iron

P01
Concerns the fine turning of steel and cast steel at high cutting speeds and small feeds. Also where close tolerance machining and good surface finish are required.

P10
Covers turning, copy turning, threading and milling of steel and cast steel at high cutting speeds with small to medium feeds.

P20
Relates to turning, copy turning and milling of steel, cast steel and long chipping malleable cast iron at medium cutting speeds and medium feeds.

P30
Covers turning and milling of steel, cast steel and long chipping malleable cast iron at medium to low cutting speeds and with medium to high feeds and also where conditions are slightly less favourable.

P40
Concerns turning of steel, steel castings with sand inclusions and cavities at low cutting speeds and large feeds and for machining under unfavourable conditions. Also for work on automatic machines.

P50
Relates to turning at low cutting speeds and large feeds of steel,

medium or low tensile strength steel castings also with sand inclusions and cavities. It also covers machining under unfavourable conditions and for work on automatic machines.

M

Colour code YELLOW

Workpiece materials which fall into this main group are:

Hard manganese steels
Austenitic steels
Cast steels
Alloyed cast irons
Nodular SG cast iron
Malleable cast iron
Non-ferrous metals

M10
Covers turning of steel, steel castings, manganese steels, grey cast iron and alloy cast iron at medium to high cutting speeds with small to medium feeds.

M20
Relates to steel, austenitic steels, manganese steel, cast steel, nodular SG cast iron and malleable cast iron in turning and milling operations at medium cutting speeds and medium feeds.

M30
Concerns turning and milling of steel, austenitic steel, heat resisting alloys, cast steel and cast iron at medium cutting speeds and medium feeds.

M40
Relates to turning, turning with form tools and parting off on lower tensile strength steels, free machining steels non-ferrous metals and light alloys especially on automatic machines.

K

Colour code RED

Workpiece materials which fall into this main group are:

Cast iron

Short chipping malleable cast iron
Hard cast iron
Non-ferrous metals
Hardened steel
Plastics
Wood
Non-metallic materials

K01
Covers fine turning, fine boring and finish turning of hardened steel, hard cast iron, high silicon aluminium alloys, abrasive plastics and other non-metallics.

K10
Concerns turning, milling, drilling and boring of grey cast iron over 200 Brinell, short chipping malleable iron, copper alloys, aluminium silicon alloys, plastics, glass, hard rubber, ceramics and stone.

K20
Relates to turning, milling, drilling and boring under conditions which require a tougher hardmetal when machining grey cast iron up to 220 Brinell, copper, brass, aluminium and other non-ferrous metals.

K30
Covers turning and milling under unfavourable conditions of low hardness grey cast iron and low tensile steel.

K40
Relates to turning of softwood or hardwood and non-ferrous metals under unfavourable conditions and with the possibility of the use of large cutting angles.

A diagramatic representation of how the ISO application group standard relates to such factors as the properties of the hardmetal required to perform the application and the features of the workpiece material is shown in Figure 41.

5.1.6 WORKPIECE MATERIALS

The iron based materials, steels and cast irons, are by far the greatest volume of workpiece materials. They also present the greatest spread of properties and of machinability.

RELEVANT FACTORS	ISO APPLICATION GROUPS
	P01 P10 P20 P30 P40
	M10 M20
	K01 K10 K20 K30 K40
PROPERTIES OF HARDMETAL REQUIRED :	
TOUGHNESS	increasing →
WEAR RESISTANCE	← increasing
WORKPIECE FACTORS :	
CHIP CROSS SECTION :	
SMALL	01 – 15
MEDIUM	05 – 25
LARGE	10 – 35
CONDITION OF WORKPIECE :	
PREMACHINED	01 – 25
AS FORGED / AS CAST	20 – 35
INTERRUPTED CUTTING	20 – 40
ABRASIVE CHARACTERISTICS :	
HIGH	01 – 20
LOW	10 – 35
'STICKY' MATERIAL :	
YES	20 – 40
NO	01 – 25
CUTTING SPEED	← increasing
FEED	increasing →
CONDITION OF MACHINE TOOL :	
LACKING RIGIDITY	20 – 40
RIGID	01 – 25

Fig. 41 ISO Application Groups – Simplified Representation

With steels, increasing the carbon content increases the hardness which in turn reduces the machinability. Additions of chromium form carbides which are hard and reduce machinability. Stainless steels have higher additions of chromium which increase their hot strength and reduces machinability. Manganese additions cause work hardening to take place and also increase the toughness of the steel and both these factors affect machinability adversely. Nickel toughens the steel and increases its tensile strength which in turn makes it difficult to machine. Molybdenum and tungsten both form hard carbides, higher contents of molybdenum and tungsten are present in high speed steel which is very difficult to machine especially in the hardened condition. Whether a steel is in the annealed condition or in a heat treated condition will obviously affect machinability.

The free machining steels contain additions of lead, sulphur and phosphorus which combine with the other elements in the steel to affect its microstructure and cause it to machine more easily.

A representation of the machinability of different types of steels and cast irons is given in Figure 42. Within each type of material there is a range of machinability. For example in the case of plain carbon steels their machinability reduces as the carbon content increases because they become harder. One further point is that there is an overlap in machinability at the lower and upper ends of most of the groups.

When choosing cutting materials the shape and surface condition of the workpiece must be considered. The workpiece may have holes or grooves or be uneven in shape which will cause interruption to the cutting during turning. Similarly the workpiece may be a casting which has a rough, uneven skin and may have sand inclusions. Both these conditions require the selection of a tougher cutting material than is needed for continuous cutting of clean metal.

The rest of this chapter discusses the popular workpiece materials used today and hardness figures, cutting speeds and feeds are quoted. These values are purely typical ones for the workpiece materials and it is not intended that the upper and lower figures represent sharp cut off points.

Aluminium and Aluminium Alloys

From the turning point of view these workpiece materials can be divided into two basic groups. The first group consists of pure or almost pure aluminium and lowly alloyed aluminium. These materials are soft

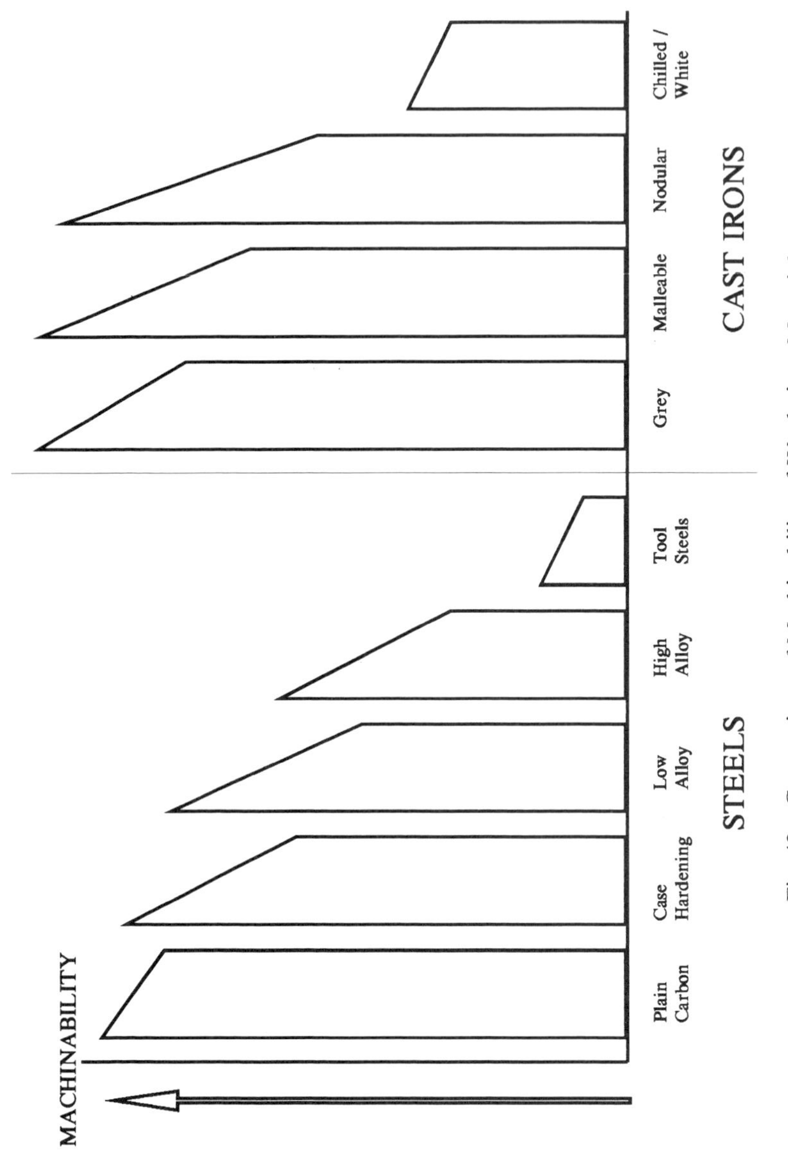

Fig. 42 Comparison of Machinability of Workpiece Materials

and tend to build up rapidly on the rake face and so a very high positive rake cutting geometry is needed for the best results. Sharp cutting edges are also an important feature. Coolants designed for machining aluminium can help to produce an excellent surface finish.

The first choice of cutting material for turning this group of soft workpiece materials is hardmetal. The hardest plain Co-WC grades are used. Special hardmetal indexable inserts are available which have very high rake cutting geometry of the order of 25° and which are ground all round giving them a very sharp edge. Cutting speeds of up to 1000 m min^{-1}. and more are commonly used at feeds ranging from 0.5 to 0.1 mm/rev.

The second group of aluminium based materials are the high silicon aluminium alloys. They are very abrasive and are comparatively tough. Once more hardmetal is a first choice of cutting material for general turning work and the same cutting geometry is used as for the simpler aluminium materials. The same feeds are also acceptable but because of the very abrasive nature of these high silicon alloys cutting speeds with hardmetal have to be reduced to around 50% of those able to be used on the simpler workpiece materials.

Much higher cutting efficiency and better surface finish can be achieved by using PCD cutting material. Because of the extremely high abrasion resistance of PCD, cutting speeds of 1000 m min^{-1}. and over are possible with the same feeds that are used for hardmetal.

Turning of aluminium and aluminium alloys falls into the ISO application group range K01 to K10.

Brass, Bronze, Copper – The Non-Ferrous Metals

These workpiece materials are easily machined and present no problems for hardmetals which are their most widely used turning material. They are soft – of the order of 100 BHN – and positive rake geometry should be employed. Cutting speeds in the range 200 to 300 m/min. are generally used for turning and typical feeds are from 0.5 to 0.1 mm/rev.

Turning of copper commutators is done with hardmetal using sharp cutting edges and positive rake geometry but PCD will perform more efficiently by operating at higher speeds, generally twice those used for hardmetal, and by maintaining its sharp cutting edge for a considerably longer time.

The hardmetals used are the 6% Co-WC alloys having a hardness of 1600 VDH and above. The ISO application group range for turning this class of workpiece material is K01 to K10.

Cast Irons

1. Grey Cast Iron

The grey cast irons cover a range of materials within a hardness band of ca. 180 to 300 BHN. The coated hardmetals are the main cutting materials used for turning grey cast iron. Typical feeds used are from 1.0 to 0.1 mm/rev. and cutting speeds are from as low as 50 m min^{-1}. for heavier work to 300 m min^{-1}. for fine finishing. Grey cast irons are abrasive and may also have inclusions and scale to be coped with. For abrasive situations, particularly on high speed finishing cuts, ceramics should be considered.

There is an increased tendency to use ceramics for turning grey cast iron in high volume production such as the automobile industry. Cutting speeds can be increased over those for hardmetal by up to two times for the same feed. This requires machines having greater power capability and also more rigidity. Such machines did not exist in earlier days and many are purpose built to carry out the operations required.

Negative rakes are usually employed with coated hardmetals and this is especially so for interrupted cutting. Neutral rakes are used for lighter general purpose work and positive rakes are preferable for fine finishing and machining thin wall section components. Grey cast iron presents no problems with difficult swarf and therefore chip control grooves are not needed.

Coolant is not normally used when machining grey cast iron but if it is felt to be necessary it should not be used with the alumina based ceramics when roughing and only used if it gives an advantage when fine finishing.

The ISO application groups which cover the turning of grey cast iron are from K10 to K40.

2. Ductile Cast Irons (Nodular SG/Malleable Iron)

For the purpose of turning, this range of cast irons can be dealt with as one group. Their general hardness spread is from 130 to 350 BHN. They are abrasive and more difficult to machine than grey cast irons. It is not easy to achieve a good surface finish with ductile cast irons. Coated hardmetal grades are the general choice for turning these workpiece materials and the feeds used are the same as those for grey cast iron. However, cutting speeds are about 20% lower than for grey cast iron, say 50 to 250 m min^{-1}.

As with grey cast irons ceramics perform well at higher cutting speeds and the same remarks apply as those made above for grey cast iron but cutting speeds are generally 10% lower.

With coated hardmetals the same cutting geometries are used as with grey cast iron. Some advantage may be gained by using appropriate chip control grooves at low feeds and small depths of cut.

The K10 to K40 ISO application groups are also the ones which apply to turning of ductile cast irons.

3. Hard Cast Iron

This group includes chilled cast iron, Ni-hard and high chromium iron which range in hardness from 67 to 90 Shore. This high hardness produces very high cutting forces when turning is being done. These hard cast irons are also very abrasive.

The alumina based ceramics and CBN are the best choice of cutting materials for repeated work but these cast irons can also be machined with the very hard 6% Co-WC grades of hardmetal with a hardness in excess of 1600 VDH. In this case a broad cut is recommended with a sharp cutting edge and neutral or small positive rake geometry. Because of the high cutting forces speeds and feeds with hardmetal are low – 15 to 30 m min^{-1}. and 0.3 to 0.1 mm/rev.

The ISO application groups which best fit these workpiece materials are K01 to K10.

Steels

1. Free Machining Steels and Low Carbon Steels

These two classes of steel can be grouped together from a machinability point of view. Their hardness falls within a range of 100 to 200 BHN. The free machining steels, as their name implies, present no problems when turning. Chip control is straightforward and high cutting speeds can be achieved with the appropriate cutting material. The low carbon steels are soft and gummy and have the tendency to form a built up edge. Chip control is a little more difficult than with free machining steels. High cutting speeds can be used and with the finishing cuts which usually go with high speeds a cermet can be a good cutting material to employ.

The most popular choice of cutting material is the range of coated hardmetal indexable inserts available. If uncoated hardmetal is chosen then the TiC containing alloys must be used or cratering will result.

Cutting speeds for uncoated hardmetal lie in a range from 60 m min^{-1}. for the heavier work and up to 200 m min^{-1}. for lighter cutting, feeds are from 1.2 to 0.1 mm/rev.

With coated hardmetals speeds are generally higher, from around 90 m min^{-1} up to 400 m min^{-1} for the finishing cuts. Feeds tend to be somewhat lower than with uncoated hardmetals at the heavier end with 0.8 mm/rev. being a good figure to have as an upper limit.

For heavy and interrupted turning, negative rake cutting geometry must be used and chip control grooves must be incorporated. Because free machining steels and low carbon steels are not high strength workpiece materials neutral or positive rake cutting geometry with appropriate chip control grooves is used for general purpose and finishing cuts.

The whole range of ISO P application groups is involved with these steel workpiece materials i.e. P01 to P50.

2. Alloy Steels and Medium to High Carbon Steels

The steels included under this heading cover a wide spread of hardness from 150 to 350 BHN. Obviously, the higher the hardness of the steel then the higher the cutting force needed to turn it. Alloying tends to increase work hardening and increasing the nickel content makes chip control more difficult. Higher carbon contents make the steels more abrasive and when combined with more alloying this has a deleterious effect on machinability.

The popular choice of cutting materials for turning these steels are the coated grades of hardmetal indexable inserts. Cutting speeds for coated hardmetal inserts range from 45 to 300 m min^{-1}. depending on the hardness of the workpiece material to be turned and the feed being used. Feeds range from 0.8 to 0.1 mm/rev.

Cutting geometries used are negative rake angles for the heavier and medium work and neutral for the lighter cuts. Chip control grooves are essential and the correct choice of groove is important.

The ISO application groups P01 to P50 include all the turning possibilities for these workpiece materials. If uncoated hardmetals are chosen to machine these steels then they must be the TiC containing grades of hardmetal which are designed to resist cratering.

Several of these workpiece materials can be heat treated to improve their hardness and one special case is that of steels used for bearings e.g. En31 where, after heat treatment, the surface to be turned has a hardness >50 HRC and traditionally has been ground to achieve the required dimensional accuracy and surface finish. CBN is now being used to turn the heat treated components and gives an excellent surface finish, quite comparable with grinding. The extreme hardness of CBN and therefore high abrasion resistance make it well able to maintain accuracy over long time cutting and floor to floor times can be considerably improved over grinding. In this case cutting speeds are of the order of 90 m min^{-1}. with feeds of from 0.3 to 0.1 mm/rev.

3. Tool Steels (Hot Work, Cold Work and High Speed Steel)
In the soft, not heat treated condition, this package of workpiece materials falls into an approximate hardness range of 150 to 250 BHN. They are abrasive materials which tend to work harden and they produce tough chips which are difficult to break.

The cutting materials which are popularly used to turn tool steels in the soft condition are the coated grades of hardmetal indexable inserts. The cutting speeds fall into a bracket of 50 to 250 m min^{-1}. and the feeds used range from 0.5 to 0.1 mm/rev.

Negative rake geometries are necessary for interrupted and roughing cuts and neutral rakes are generally accepted for the lighter turning operations. Chipgrooves are needed and should be selected to be suitable for the turning operation.

The components made from these materials will be heat treated after these turning operations have been carried out and so no accurate fine finishing is needed at this stage and in any case some distortion may occur during heat treatment. Thus the ISO application groups which cover these steels in the soft condition are the P20 to P30 range.

When heat treated these steels become very hard and fall into a range from 55 to 65 HRC. This now makes then much more difficult to machine. The preferred cutting material from a technical point of view is CBN but ceramics also perform and are cheaper although their life is considerably shorter.

For hardened tool steels the recommended cutting speed for CBN lies in the range 50 to 120 m min^{-1}. with a spread of feed from 0.5 to 0.1 mm/rev. Speeds and feeds for ceramics are of the same order.

The ISO application groups covering the turning of these hardened tool steels is from K01 to K10. The very hard, plain Co-WC grades of hardmetal can be used without any cratering problems. Neutral to positive geometries with sharp cutting edges and low cutting speeds are necessary but their life is considerably shorter than that of CBN or ceramics.

4. Austenitic Stainless Steel
The hardness of austenitic stainless steel workpiece materials lies in the general region of 135 to 275 BHN. Their high nickel and high chromium

content gives them some problems in turning. They are subject to rapid work hardening and are very difficult to machine with small depths of cut. The chips which are produced are tough and stringy and chip control requires the selection of indexable inserts with chip control grooves specifically designed for these materials. Although austenitic stainless steels are not very hard they are very abrasive.

For general turning work the range of coated hardmetal indexable inserts is the most popular cutting material. The cutting speeds are from 75 to 220 m min^{-1} with feeds of 0.8 to 0.1 mm/rev. Because of the abrasive characteristics of austenitic stainless steels cermets can be advantageous when lighter cuts are being taken. In the case of cermets speeds can be somewhat higher, 150 to 280 m min^{-1} with feeds of 0.4 to 0.1 mm/rev. being typical.

Austenitic stainless steels fall into the M20 to M30 ISO application groups. If interrupted or heavier cutting is involved then negative rakes are needed. With lighter cutting, neutral or positive rakes can be employed. In all cases chip control grooves are necessary.

If uncoated hardmetals are used for turning no crater will be formed when cutting austenitic stainless steels and so the plain Co-WC hardmetals can be used. Cutting speeds will then be lower than with coated hardmetals.

5. Ferritic and Martensitic Stainless Steels

These workpiece materials fall into two groups of hardness. One is in the range 130 to 300 BHN and the other from 300 to 450 BHN. They are both characterised by high work hardening and they produce brittle and stringy chips. High cutting forces are generated when turning them.

Coated hardmetal indexable inserts are the popular choice for cutting these alloys and speeds range from 45 to 150 m min^{-1}. The feeds used are from 0.5 to 0.1 mm/rev. Negative rake geometry is the choice for heavier and interrupted cutting with neutral and positive rakes for the lighter and finishing cuts. Chip control grooves must be incorporated for satisfactory swarf control.

As with austenitic stainless steels cermets can be used for the lighter cutting. Speeds commence at 150 m min^{-1} for both hardness ranges and go up to 250 m min^{-1} for the softer group of steels and 220 m min^{-1} for the harder group. In both cases feeds range from 0.4 to 0.1 mm/rev.

Uncoated hardmetals can be used with similar cutting geometry to that of the coated hardmetals. Cutting speeds are somewhat lower but the feeds are the same. Because these ferritic and martensitic stainless steels do produce a crater the uncoated hardmetals must be chosen from the TiC containing alloys.

The ISO application group band for these workpiece materials is P10 to P30.

6. Heat Resisting Alloys (Iron, Nickel and Cobalt-Based)

The main property of this class of metals is that they are still very strong when they are hot. This makes them one of the most difficult group of workpiece materials to machine and all the cutting materials used to turn them show relatively poor tool life when compared with turning other workpiece materials.

They are prone to rapid work hardening and small depths of cut are difficult to carry out. The chips which they produce are tough and stringy and chip control presents a problem. Although they are not excessively hard they are very abrasive. All these factors combine to make them one of the hardest tasks for the cutting tool engineers. For machining purposes it is possible to divide them into two hardness groups – 125 to 250 BHN and 200 to 450 BHN.

Coated hardmetal indexable inserts are probably the most used cutting material. With the lower hardness heat resisting alloys coated hardmetals operate in a speed range of 20 to 100 m min^{-1} and with the higher hardness alloys from 20 to 50 m min^{-1}. In both cases feeds span from 0.3 to 0.1 mm/rev.

Sialons, with their very high hot hardness, have proved to be successful in turning these alloys. They do not like interrupted cuts and should be employed on clean metal. Cutting speeds for the lower hardness workpiece materials are from 120 to 230 m min^{-1} and for the harder alloys are from 90 to 215 m min^{-1}. Again feeds are from 0.3 to 0.1 mm/rev.

CBN is becoming a choice of cutting material for the harder alloys and typical cutting speeds range from 100 to 160 m min^{-1} at an average feed of about 0.2 mm/rev.

Cermets are used on the lower hardness workpiece materials at cutting speeds of 30 to 150 m min^{-1} and feeds of 0.3 to 0.1 mm/rev.

Uncoated hardmetals can also be employed but probably on the lower hardness group of alloys with reduced cutting speeds of from 15 to 60 m min^{-1} and feeds of 0.8 to 0.1 mm/rev. The Co-WC alloys are used as no cratering will occur.

With both coated and uncoated hardmetals negative rake geometry is needed with appropriate chip control grooves.

The whisker reinforced alumina based ceramics are showing some success in cutting heat resisting alloys. Round indexable inserts are preferred and comparatively high cutting speeds of 150 to 350 m min^{-1} with average feeds of 0.2 mm/rev. are being employed.

The M20 to M30 ISO application groups cover the most popular turning tasks for these workpiece alloys.

7. Titanium and Titanium Alloys

As with heat resisting alloys, titanium based workpiece materials are difficult to machine. They are abrasive, their chips are tough and stringy and tend to gall and weld to the cutting edge. Cutting speeds are necessarily low when compared with steels of similar hardness and tool life is relatively poor. They also tend to work harden and are prone to produce a glazed surface. The main cutting materials which are chosen to turn them are the uncoated and coated grades of hardmetal indexable inserts. Positive and neutral rakes are the preferred cutting geometries with selected chip control grooves. Sharp cutting edges should be employed and coolants can be advantageous.

The titanium based workpiece materials can be divided into three hardness groups for the purpose of comparing cutting speeds when turning. These are 100 to 200 BHN, 250 to 350 BHN and 350 to 400 BHN.

With uncoated hardmetals the cutting speeds corresponding to these hardness groups are 60 to 125 m min^{-1}, 30 to 75 m min^{-1} and 10 to 50 m min^{-1} respectively. For coated hardmetals they are 60 to 150 m min^{-1}, 30 to 100 m min^{-1} and 10 to 60 m min^{-1}. For both types of hardmetal the spread of feed is from 0.4 to 0.1 mm/rev.

No cratering occurs with titanium and its alloys and so the 6% Co-WC plain hardmetal grades are preferred when uncoated hardmetal cutting material is used.

The ISO application group K20 best fits this class of workpiece materials.

8. Plastics, Nylon, Hard Rubber and Similar Non-Metallics

These materials cannot be described as having high strength in relation to those discussed above. They do not shear in the same way or produce long chips when being turned as is the case with most steels. Positive cutting geometry is necessary together with sharp cutting edges so that the push off force is at a minimum. Uncoated hardmetals are generally used for turning them and the same cutting geometry used for aluminium and its alloys with the same sharp edge condition is the best choice. Because they do not give rise to high cutting forces and by using positive geometry they can be turned at very high speeds without generating high cutting temperatures. However these high speeds demand that the cutting materials used must have good abrasion resistance. The hardest Co-WC grades of uncoated hardmetal are used and even with the much reduced wedge angle resulting from the high positive cutting geometry there is little danger of breakage. For repeated production on some of the carbon composites and similar non-metallic materials polycrystalline diamond is used at cutting speeds around 1000 m min^{-1} and feeds averaging 0.2 mm/rev. Cutting speeds for uncoated hardmetal range from 200 to 600 m min^{-1} with feeds of 0.5 to 0.1 mm/rev.

This group of workpiece materials is classified as K01 to K10 in the ISO application system.

Comment

The workpiece materials listed cover those used for the majority of components produced by turning. The breakdown of cutting materials employed to machine these components by turning falls into an approximate breakdown of:

70% HARDMETAL
10% HIGH SPEED STEEL
20% OTHERS

80% of the indexable inserts which are included in the hardmetal grouping are now coated.

Although high speed steel represents some 10% of the cutting materials used for turning it is not the most popular choice for machining any of the workpiece materials described. Its limitations show up as the workpiece materials increase in hardness. However, there are cases where it has a strong foothold in the market. These are essentially where speeds are limited and where heavy cutting is not practical. Such cases are machine tool orientated. The small lathes found in many workshops have very limited power and rigidity and will not operate at high speed. These machines are totally unsuitable for hardmetal and high speed steel tool bits ground by the operator are the popular cutting material to use.

High speed steel dove tail and circular form tools are also the main choice of cutting material on multi spindle automatic machines such as

Which Cutting Material?

Workpiece material	HSS	Stellite	HM	Cermets	Al_2O_3 ceramic	Si based ceramic	CBN	PCD
Aluminium and Al alloys	Y		Y					Y
Brass, bronze, copper and non-ferrous metals	Y		Y					Y
Grey cast iron			Y		Y	Y		N
Ductile cast iron			Y		Y			N
Hard cast iron	Y		Y			N		N
Free machining steels and low carbon steels	Y		Y			N		N
Alloy steels and medium to high carbon steels			Y			N		N
Tool steels: Soft condition			Y			N		N
Hard condition	N	N	Y		Y		Y	N
Austenetic stainless steels			Y	Y				N
Ferritic and martensic stainless steels			Y			N		N
Heat resisting alloys			Y			Y		N
Titanium and Ti alloys			Y					
Plastics and non-metallics			Y					Y
Hardmetal	N	N	N	N	N	N	Y	Y

those used in the bearing industry. The additional loads on the machines which result from the use of hardmetal cutting tools quickly show up their lack of rigidity and so high speed steel is preferred.

The facing table gives a general idea of the cutting materials used to turn the main variety of workpiece materials. Where the letter 'Y' is used it is definitely possible to use the designated cutting material. Where an 'N' is used one should not attempt to turn with that cutting material. Where no letter is given then this does not mean it is not possible to use that cutting material but it may not be the best, or even a sensible choice. For example using CBN on soft workpiece materials would offer no advantage whatsoever and would be too expensive to employ. On the other hand high speed steel will turn grey cast iron but unless the lathe used has very low power or is very unstable then the hardmetals and ceramics will be vastly superior in performance and must be preferred to high speed steel.

5.1.7 COOLANTS

One fact that must be accepted with turning is that it is not possible to transfer a continuous flow of coolant to the actual cutting edge. Thus the primary action of a coolant is not to bring down the temperature of the cutting edge and so increase the life of the cutting tool. Indeed many turning operations are carried out without the use of coolant.

With some cutting materials coolants cause thermal shock problems which result in cracking of the tool. When using Al_2O_3 ceramics to turn steel, coolant should not be employed because of the thermal shock situation. Similarly, when the white Al_2O_3 ceramics are used to turn cast iron coolant should be avoided.

Cermets are more sensitive to shock than hardmetals and if rough turning is being carried out with cermets then coolant is not recommended. However, with very light finishing operations coolant can be beneficial to the surface finish of the workpiece.

The cutting action of CBN is to raise the temperature of the workpiece to a point at which it softens and can then more easily be machined and so the use of coolant is exactly not what is required.

Coolants can help to produce a good surface finish on the workpiece. This is particularly so with softer materials e.g. low carbon and free machining steels and also with aluminium and its alloys where coolants exist which are specially formulated for use on aluminium.

All the cutting tool manufacturers strongly recommend the use of coolant when machining heat resisting alloys and also for titanium and its alloys.

Coolants can help with swarf removal. By positioning the flow of coolant in the appropriate direction the chips can be flushed away from the working area and reduce the possibility of them fouling up.

Finally, by using coolant, the workpiece itself may be kept down to a temperature which is satisfactory for handling by the operator when the turning process is finished.

In case of doubt, or for more detailed information, it is recommended that contact be made with the major cutting tool suppliers who have excellent technical back up services.

5.1.8 TYPES OF FAILURE OF HARDMETAL CUTTING TOOLS

Because hardmetal is the most important of all the cutting materials for turning operations and also because the range of alloys available offers the possibility to change the grade being used to overcome the failure problem this chapter will only deal with typical failures which can occur with hardmetal tools.

Crater Formed Behind the Cutting Edge

This mode of failure should only be experienced when turning with uncoated hardmetal tooling. The way in which a crater is formed has been fully covered in 2.3 and an illustration of a crater is shown in Figure 43. If cratering becomes excessive it will break through the cutting edge and immediate failure will occur.

The following corrective action can be taken:

a) Change to a cutting material which is more resistant to the formation of a crater. This could be a coated hardmetal or an Al_2O_3 ceramic.
b) Try a more positive rake geometry if conditions are suitable.
c) Reduce the temperature being generated at the cutting edge by lowering the cutting speed. Further advantage may be gained by reducing the feed.

Fig. 43 Crater Formation in Hardmetal

If no other geometry or cutting material is available then reducing the speed sufficiently will solve the problem of crater. Even with uncoated plain Co-WC hardmetal grades of cutting material no crater will be formed when turning ferritic steels if the cutting speed is kept below 40 m min^{-1}.

Plastic Deformation of the Cutting Corner

This problem is normally only encountered with hardmetal and in particular with hardmetal indexable inserts. It occurs when excessive temperature and pressure cause the nose of the cutting tool to plastically deform. The cutting edge geometry is then destroyed and the operation fails.

Two solutions to the problem are available. The first is to use a harder grade of cutting material which will have better resistance to plastic deformation. The second is to lower the temperature by reducing the cutting speed and/or reduce the feed.

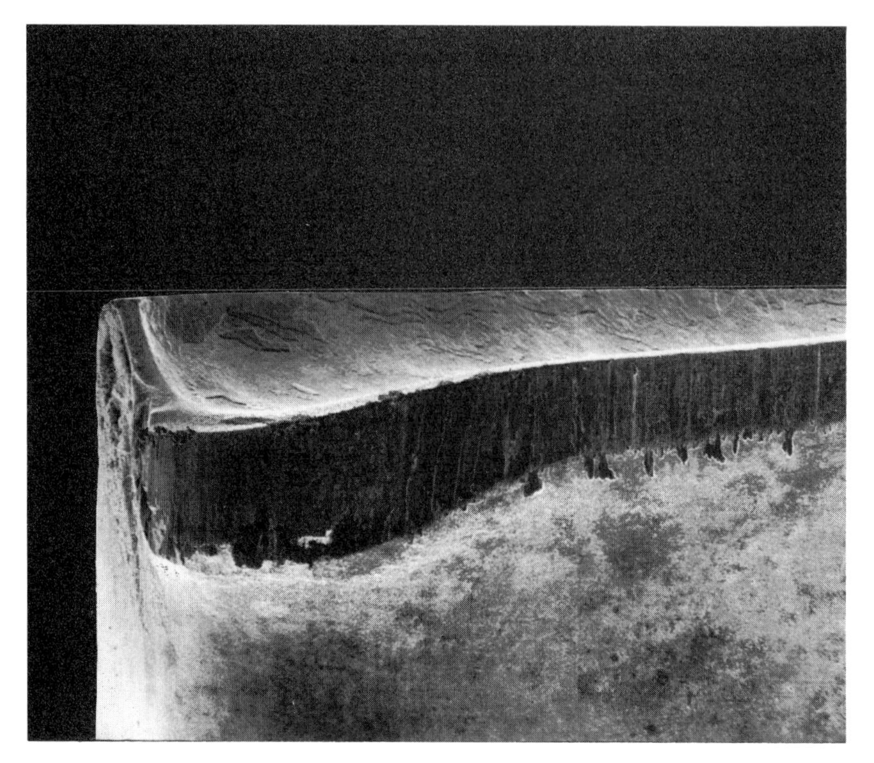

Fig. 44 Plastic Deformation in Hardmetal

Figure 44 shows the nose of an indexable insert which has been plastically deformed. The cutting edge would normally form a straight line all round the nose but in this case it has a considerable droop.

Clearance Face Wear

As the wear on the clearance face of an insert increases it causes the cutting forces to increase which also brings an increase in temperature of the indexable insert. There is a greater tendency for vibration to occur and there is a reduction in the quality of the surface finish of the workpiece being machined. There is also a deterioration in the ability to keep to size on the machined component. An illustration of a cutting edge where clearance face wear is approaching its acceptable limit is shown in Figure 45.

If excessive clearance face wear is occurring it can be combated by reducing the cutting speed. Alternatively a harder grade of hardmetal

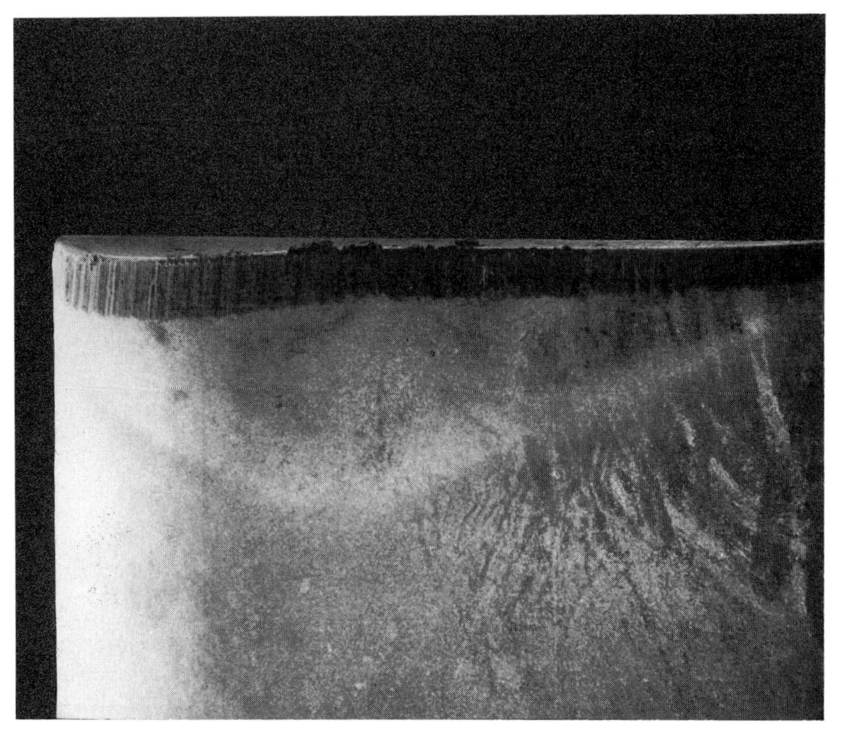

Fig. 45 Clearance Face Wear in Hardmetal

may be used but care should be taken that its toughness is sufficient to stand up to the task or breakage will be the result which is even more costly than clearance face wear. If a low feed rate is being used then increasing the feed rate will tend to reduce the clearance face wear for a given metal removal rate.

Chipping and Notching

Chipping and notching generally indicate that there is excessive mechanical loading on the indexable insert. Figure 46 shows a photograph of a cutting edge which is suffering from this problem.

One cause may be vibration or another cause could be that the feed and depth of cut may be greater than is desirable. It could also be that the grade of hardmetal being used is too brittle. Interrupted cuts can also be a reason for chipping and notching.

To overcome vibration selecting an insert with a more positive rake

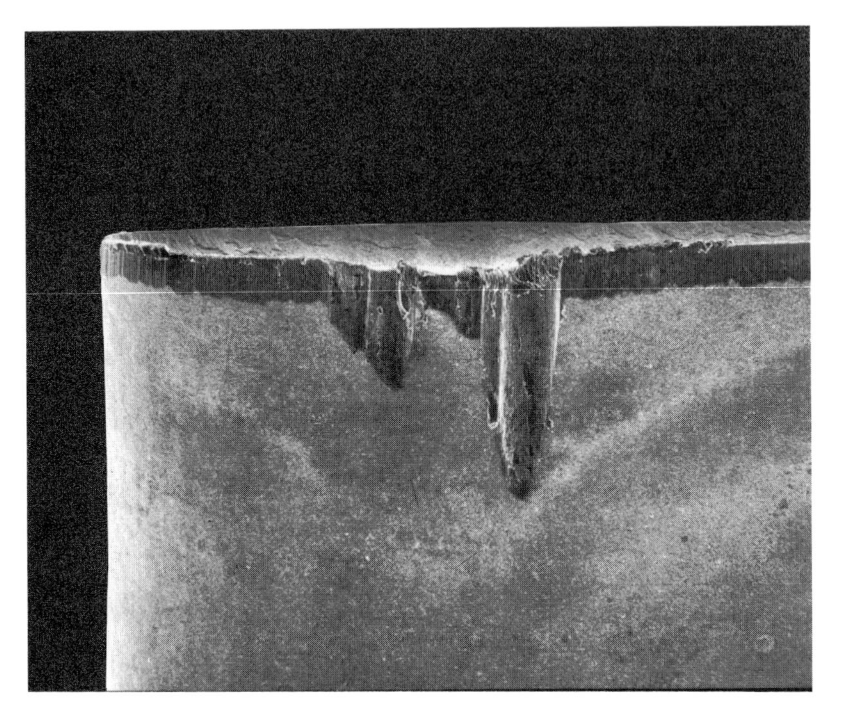

Fig. 46 Chipping and Notching of Hardmetal

could help. Using a smaller nose radius will also lower the push off force. Ensure that the tool is cutting at the correct centre height and make sure that the overhang of the tool is at a minimum.

If it is felt that the cause relates to feed and depth of cut then the feed should be reduced but if it is desired to maintain the feed and depth of cut then a tougher hardmetal grade would help. Negative rake and a strong edge condition should also be employed.

If the problem relates to interrupted cutting then use a strong edge condition together with negative rake and if failure still occurs then use a tougher grade of hardmetal. Also ensure that the turning operation is being carried out with the best possible stability (machine and holding devices).

It is assumed that a chip control groove which is appropriate for the turning operation is being used.

Built-Up Edge

Built-up edge has already been described in chapter 5.1.2 and is illustrated in Figure 34. Solutions to the problem of built up edge are also given in that chapter.

A diagrammatic representation of the most popular solutions to some of the failure problems which occur when turning with hardmetal is given in Figure 47.

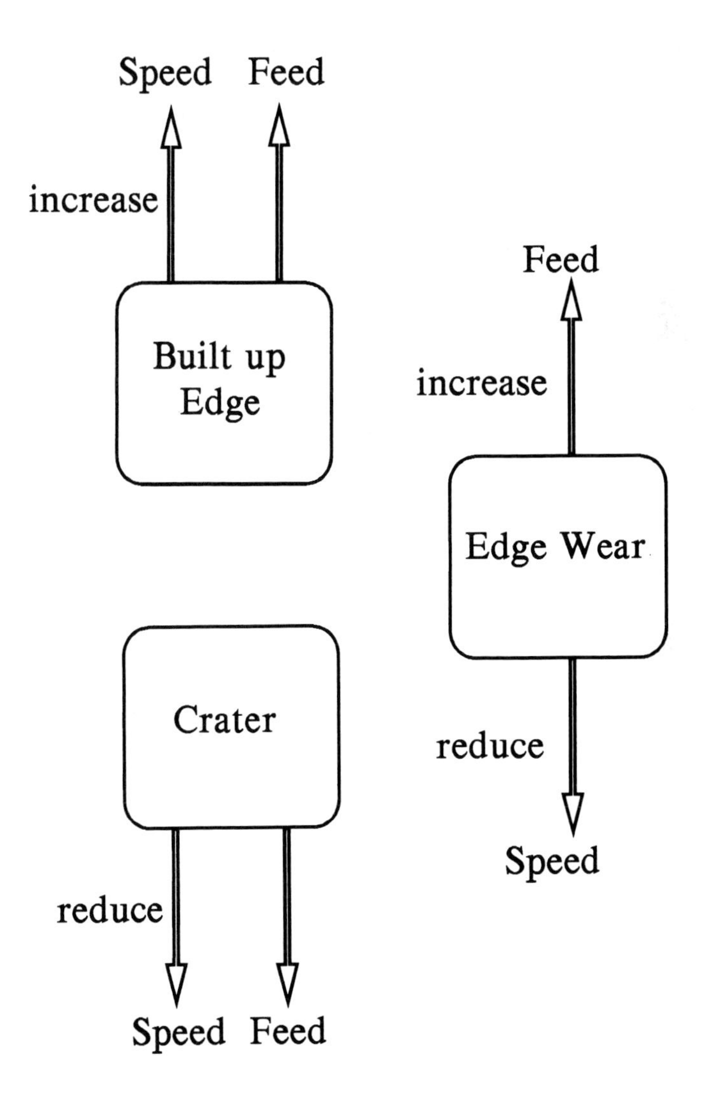

Fig. 47 Problem solving when Turning with Hardmetal

5.1.9 TURNING TOOLS

Turning tools can be divided into three main groups. These are:

a) High Speed Steel tools which consist of tool bits, butt welded tools, tipped tools and form tools.
b) Brazed hardmetal tools.
c) Toolholders which carry indexable inserts.

High-Speed Steel Tools

High-speed steel single point turning tools are available in three forms. These are either solid pieces of high speed steel, usually for smaller tools or secondly, butt welded tools where the head is made from high speed steel which is then welded to a less expensive medium carbon steel shank. Butt welded tools are in the medium size range. The third possibility covers very large tools where the tip of the tool is a piece of high speed steel which is brazed onto a steel shank. In this latter case the braze material has a high melting point which makes heat treatment possible without the tip becoming detached. Both the butt welded and the tipped high speed steel tools are heat treated after the joining operation. The cutting profiles are roughed out whilst the tools are in the soft condition and so final grinding after heat treating is minimised.

Solid high speed steel tool bits are specified in a British Standard – BS 1296 : Part 4. This is essentially based on ISO 5421 with only very slight differences in the tolerances for certain sections and on lengths. These ground tool bits cover three cross sectional shapes, these are round, square and rectangular.

For the round tool bits eight diameters are specified ranging from 4 to 20 mm and five lengths are stated beginning with 63 mm and going up to 200 mm. The longer lengths do not apply to the smaller diameters and the shorter lengths are not applicable to the larger diameters.

The square section ground tool bits have a range from 4 × 4 up to 25 × 25 mm and the same lengths are used as those for the round section tools with the remarks about lengths also applying.

In the case of rectangular section tool bits seven widths of section are specified. These are 4,5,6,8,10,12 and 16 mm. Except for the 16 mm width, two possible heights are quoted for each width e.g. 4 × 6 and 4 × 8, 5 × 8 and 5 × 10 mm. The heights range from 6 to 25 mm. Only three lengths are specified, 100, 160 and 200 mm. The 100 mm length applies

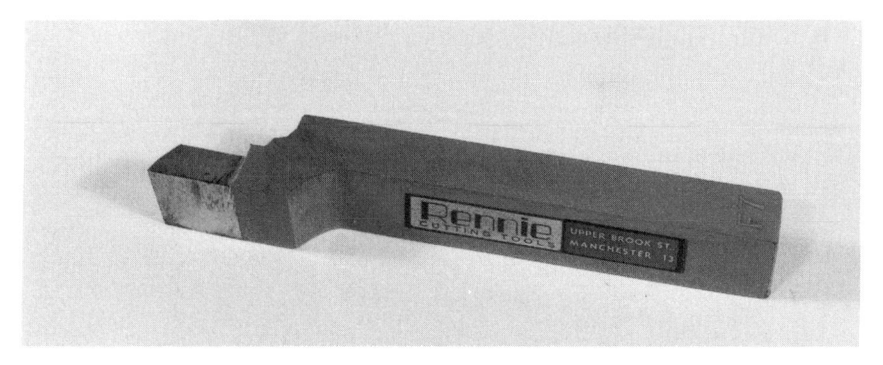

Fig. 48 Butt welded High Speed Steel Turning Tool

only to the 4 and 5 mm widths whilst the longer lengths apply to the larger sections.

These tool bits are ideal for use on small lathes and can easily be ground into cutting profiles to suit most turning applications. The 8, 10 and 12 mm square section tool bits are probably the most popular ones used in the UK.

Butt welded high speed steel tools originated in the early 1930s and are still in use today. They are specified in a British Standard BS 1296 : Part 3. As well as turning tools this standard also specifies tools for boring, shaping and planing. It also includes butt welded blanks which can be ground as required. There is also a DIN standard which specifies tool shapes.

All dimensions in BS 1296 are quoted in millimetres but similar tools exist which are dimensioned in inches and these are still very popular. A butt welded tool is illustrated in Figure 48.

Fifteen different butt welded tool profiles are covered by the British standard and tables of dimensions are given for both preferred and non preferred sizes. As well as turning tools these profiles include boring, parting off, screw cutting, planing and also recessing tools. Each tool shape has been given a reference number which is the same number by which it had been recognised prior to the issue of the standard in 1978. This reference number designates the tool shape.

The fifteen profiles listed are:

1. Light turning and facing tool – Ref. No. 1 (right hand)
2 (left hand)
2. Straight nosed roughing tool – Ref. No. 3 (RH)
4 (LH)

3. Knife tool or side-cutting tool – Ref. No. 7 (RH)
   8 (LH)
4. External screw cutting tool – Ref. No. 13
5. Parting-off tool – Ref. No. 16RH
   16LH
6. Round nose planing or shaping tool – Ref. No. 17
7. Facing tool – Ref. No. 19 (RH)
   20 (LH)
8. Right-angle recessing tool – Ref. No. 25 (RH)
   26 (LH)
9. Right-angle parting-off tool – Ref. No. 27 (RH)
   28 (LH)
10. Square nosed turning and facing tool– Ref. No. 29 (RH)
    30 (LH)
11. Cranked turning or recessing tool – Ref. No. 39 (RH)
    40 (LH)
12. Hardened blank – Ref. No. 47
13. Boring tool – Ref. No. 50 square nose
    50(A) Vee nose for internal screw cutting
    50(B) round nose
14. Swan-necked finishing tool – Ref. No. 52
15. Hardened blank – Ref. No. 62

Except for the hardened blanks, butt-welded tools are supplied with a ground cutting profile and a flat base. Chipbreakers are rarely if ever necessary with these high speed steel tools and the rake angles which are already built into the cutting geometry are those which would normally be applied in most turning operations. Consequently, with correct use, regrinding of the tool to bring it back to its original condition should only be necessary around the clearance face. Aluminium oxide grinding wheels are normally used to carry out any grinding.

High speed steel form tools are usually used on automatic machines and as their name implies they have a form which is ground into the cutting edge and this is reproduced on the component during turning.

There are three basic types of form tools the first of which have a square or rectangular shank with a butt welded head into which the form is ground.

The second are the so called circular form tools which are thick discs of solid high speed steel where the form is ground into the circumference of the disc and the rake angle is cut radially towards the centre of the disc. Circular form tools are mounted onto the machine by means of a hole through the centre of the tool and therefore can be rotated to bring the cutting edge to the correct position. Regrinding is a simple operation and consists of surface grinding the rake face to remove the clearance face wear which has taken place – the form is not touched.

The third type of form tool is presented tangentially to the workpiece and is mounted by means of a dove tail shaped projection at the back of the tool, hence they are called dove tail form tools. This system enables the tool to be moved up and down and to be adjusted by means of a screw which is located at the base of the tool holder for setting purposes. It also offers a large surface area for good rigid clamping to be effected. The form is ground down the full length of the clearance face of the tool and the top of the tool, which is the rake face, is ground flat to the rake angle required. Regrinding is done in the same way as with circular form tools, the rake face being ground down until the wear on the clearance face has been removed.

Form tools are widely used in the bearing industry on single spindle and multi spindle bar automatic machines.

Clamped and indexable high speed steel tooling has already been referred to in 2.1 and examples of form tools are illustrated in Figure 1.

Stellite Turning Tools

Stellite turning tools are available as tool bits and as tipped tools with cutting profiles similar to those of the high speed steel butt welded tools. The tool bits are solid stellite and the tipped tools are made from cast stellite tips which are brazed onto steel shanks in the same way that hardmetal tools are manufactured.

The tipped tools are supplied with the cutting profiles and rakes already ground and ready for use on the machine.

Brazed Hardmetal Tools

The situation with regard to standardisation of brazed hardmetal turning tools in the UK is not an orderly one. Originally there were two standards, the first being a British industry standard based on imperial dimensions. In parallel with this there was an industry standard for the hardmetal tips which were brazed onto the steel shanks to make the tools.

An ISO standard also existed which was updated in 1975 and is still in use today (ISO 243). The hardmetal tips used on these tools are also standardised but are different from the ones used on the British industry standard tools.

In the mid 70s the British Hardmetal Association prepared a metric tool standard which was very near to the ISO standard and used different tips from those used in the imperial standard. The intention was to replace the imperial standard by the metric standard. This would move towards ultimate compatibility with the ISO standard.

The customers were not converted to the new metric standard and we are left with all three brazed tool ranges available from tool manufacturers. Figure 49 shows a brazed hardmetal cranked turning tool.

The updated ISO 243 covers 6 external turning tools and 1 parting off tool, they are identified as No.1 No.2 etc. It specifies the hardmetal tip used and the height, the width and the overall length of the tool. Apart from tool number 4 each of the types can be provided as a left or as a right hand tool.

To identify whether an external turning tool is left or right hand cutting place the tool so that its nose is towards you and its back is away from you. Now look down on the tool. If the cutting edge is to the right then it is right hand cutting. If the cutting edge is to the left it is left hand cutting.

Most brazed turning tools are identified according to the ISO applica-

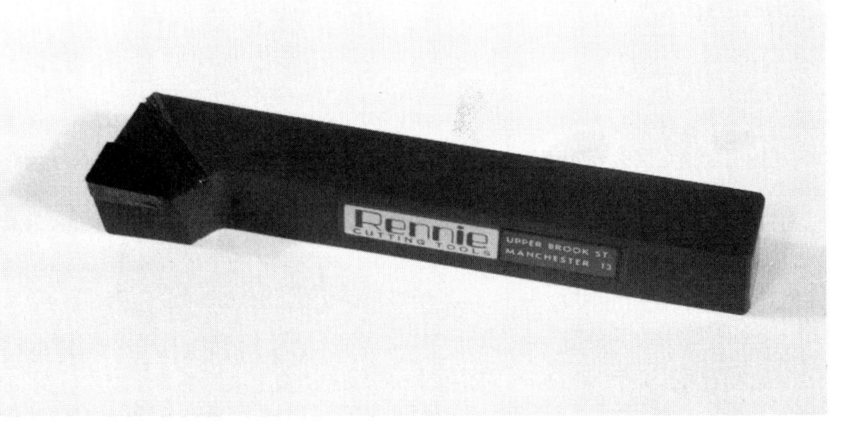

Fig. 49 Brazed Hardmetal Turning Tool

tion system. The tools to be used for turning cast iron and non-ferrous materials are painted red. The tools provided for the ISO 'M' group of applications are painted yellow. Those for cutting ferritic steels are painted blue. The application group for which the tool is intended is marked on the tool e.g. K20, P30 etc.

At some stage the question will be asked 'Should I use a brazed hardmetal tool or an indexable insert tool ?'. There is no doubt that indexable insert tools are the first choice in situations where productivity is the criterion. However, where initial cost is a critical factor such as in one-off situations the following points are relevant.

The cost of a brazed tool is about one fifth that of an indexable insert toolholder. The brazed tool can be reground between 20 and 30 times. Each indexable insert costs about half that of a brazed tool and can be indexed from 2 to 8 times depending on its shape and whether it is inclined negatively in the toolholder or not.

Perhaps the most important factor is the choice of cutting material which is available with brazed tools and that which is available as indexable inserts. With brazed tools the choice is restricted to the range of uncoated hardmetals on offer. With indexable inserts the choice is vast – from uncoated hardmetals to coated hardmetals and from there to cermets and ceramics and ultimately to cubic boron nitride and polycrystalline diamond. A further point, especially when machining steels, is the availability of highly sophisticated chip control grooves which exist with indexable inserts. The chipbreakers which are ground on brazed tools are more basic and left much to the operator's fancy.

Grinding in of chipbreakers is one of the most sensitive areas for cracking of the hardmetal and care should be taken when carrying out this operation. The grinding of brazed hardmetal tools is usually done by hand. The normal practice is to use soft, green grit silicon carbide wheels. Diamond wheels are the ideal but are much more expensive and are not the popular choice.

Special brazed hardmetal turning tools to a customer's specific requirement are sometimes needed and a service to this effect is offered by many of the toolmakers in the UK. Form tools are a good example of this point.

Tools with Indexable Inserts

This is by far the most important category of turning tooling. At least 90% of all turning tools in use today are of the clamped indexable insert type. ISO 5608 is a designation system for 'Turning and Copying Tool Holders and Cartridges for Indexable Inserts'. This is a letter and number system similar to that for the indexable inserts themselves. There are 10 positions which together provide the designation for a particular tool.

These positions are:

Position 1. (letter) – Clamping Method

Four methods of clamping are covered. The letter 'S' is used to designate the holding down of an indexable insert by a screw through a centre hole in the insert. 'M' is a combination of clamping both by means of a hole in the insert together with a top clamp. The letter 'P' designates clamping using a hole in the insert (i.e. using side pressure). 'C' is clamping from the top which applies mainly to inserts without a hole.

Position 2. (letter) – Insert Shape

A letter is used to designate the shape of the insert which is carried by the toolholder. e.g. if the toolholder takes a triangular insert then the letter used is 'T'. 'S' describes a toolholder which takes a square insert and so on. The letters used are those which are described in 3.2, Detail 1.

Position 3. (letter) – Style of the Toolholder

Twenty two possibilities exist for this position and twenty two single letters are used. Each possibility is the way in which the tool is presented to the workpiece i.e. a 90° approach angle tool for turning to a square shoulder has the letter 'A' or a 90° facing tool has the letter 'C'.

Position 4. (letter) – 'Clearance Angle' of the Indexable Insert

Although this position is termed clearance angle it does not refer to the actual clearance angle which will exist when the insert is mounted in the toolholder but is 90° minus the angle which the side of the insert makes with its top face. For example an insert with a right angled side has a 'clearance angle' of 0° and is designated by the letter 'N'.

The letters used and the principle of the system is identical with that for designating the indexable insert and is described in 3.2, Detail 2.

Position 5. (letter) – Hand of the Tool

The tool can either be right hand cutting – 'R', left hand cutting – 'L' or able to cut in either direction – 'N'.

Position 6. (two digit number) – Height of the Tool Shank
The two digit number indicates the height of the tool shank in mm. In the case of a round tool the two digits used are '00'.

Position 7. (two digit number) – Width of the Tool Shank
Again the two digit number indicates the width of the tool in mm. In the case of a round tool the two digits are the diameter of the tool in mm.

Position 8. (letter) – Total length of the Tool
Lengths are designated ranging from 32 to 500 mm. For example a length of 150 mm. is designated 'M'. The letter 'X' is reserved to indicate that the length is special and does not conform to one of the fixed lengths laid down.

The length is defined as the total length of the tool with the insert in position.

Position 9. (two digit number) – Cutting Edge Length
This position defines the size of the insert being used by reference to its cutting edge length. It is the same two digit number used in chapter 3.2, Detail 5.

Position 10. (letter) – Special Style
Position 10 is optional and can be used to define such things as the 'qualification' of precision tools.

An example of the use of this toolholder designation system is given below:

Example: PSBNR 3225P12

P = a lever locking system toolholder
S = Toolholder taking a square insert
B = A turning tool with a 75° approach angle
N = Insert with a right angled side – negative insert
R = Right hand cutting
32 = Tool shank height is 32 mm
25 = Tool shank width is 25 mm
P = Overall length of tool is 170 mm
12 = The insert is 12.5 mm (½″) square

Important

The ISO designation systems for inserts, toolholders etc. are not ordering descriptions. For example the indexable insert thickness is not specified in this toolholder designation. Hardmetal inserts, ceramic inserts and CBN inserts are all offered in differing thicknesses for several of the popular insert shapes and cutting edge lengths. Obviously the correct toolholder to carry a particular insert thickness must be selected and it is necessary to refer to the manufacturers literature or to make contact with them to obtain the correct ordering code to pair up with the indexable insert to be used.

The Toolholder Insert Pocket

The pocket in which the indexable insert sits is a crucial part of the toolholder design. Ideally the insert should be supported by two abutment faces in the pocket but this is not always possible especially with triangular inserts. With round inserts the abutment is a continuous curve.

The clamping system used should direct the insert into the two abutment faces and on locking up should pull back the insert against both faces. This then ensures correct positioning of the insert for the turning operation but more important it gives the insert the maximum possible rigidity during machining.

The seating of the insert in the pocket is also vital and where possible a flat support pad of hardmetal is located in the base of the pocket. The insert then sits on this flat bed which does not give as the insert takes the load of the main cutting force. A second feature of the hardmetal support pad is that should the insert break whilst the tool is cutting then the support pad, which is usually made from a tough grade of hardmetal, can probably withstand the shock load and prevent the nose of the toolholder from being damaged.

It is very important to ensure that the pocket is cleaned when indexing or changing an insert. Debris left in the pocket can result in incorrect seating and a great tendency for breakage of the insert. Should the pocket or nose of the toolholder become damaged it is strongly recommended that no attempt is made to patch it up by welding on new metal and 'fettling up' the pocket. Apart from the difficulty of reforming the pocket the steel used for making the toolholders is carefully selected to give very high strength when heat treated and welded metal will not have the desired strength to support the insert.

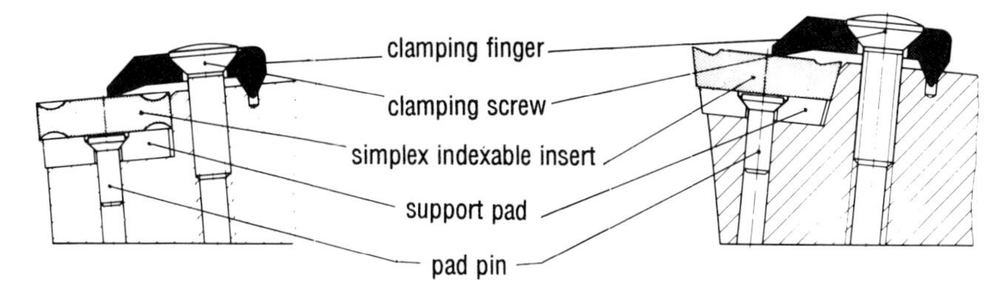

Fig. 50 Overhead Clamping of an Indexable Insert

Clamping Systems

Overhead Clamping

Figure 50 illustrates an overhead clamping system used for either negative or positive rake inserts and both cases are shown. The hardmetal support pad at the base of the pocket is held permanently in position by means of a pin which is fitted into the toolholder. The insert is then placed in position and the clamping screw is tightened. The clamp rocks forward by pivoting on its rear and the insert is therefore kept back in position in the pocket.

This clamping system is used to hold inserts which do not have a hole in them. It is suitable for light to medium turning where the cutting pressure pushes the insert back into the seat (conventional turning and facing). It is not suitable for heavy roughing operations and is not recommended for outfacing where there will be a tendency to pull the insert from the holder.

The system is designated 'C' in the ISO system.

Lever Lock Clamping

A lever operated clamping system is illustrated in Figure 51. It is only

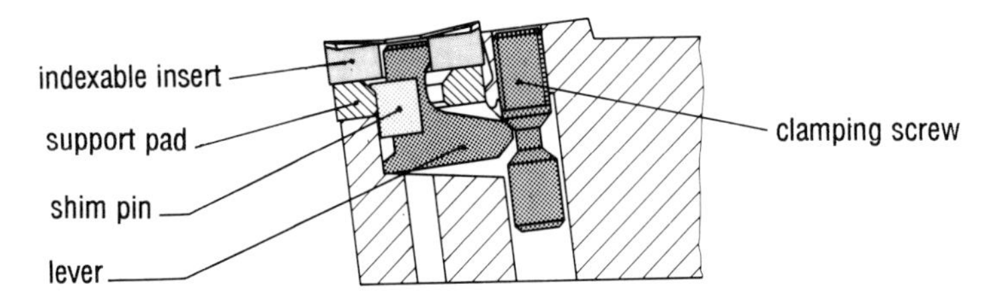

Fig. 51 Lever Lock Clamping of an Indexable Insert

suitable for negative rake inserts which have a centre hole. Positive rake inserts will tend to rise out of the pocket as the lever is pulled back.

The support pad is held permanently in position by means of a hollow pin which is pressed into a fixing hole in the toolholder. The diameter of the bore of this pin is sufficiently large to allow clearance for the movement of the lever to operate. As the clamping screw is tightened the lever pivots on its heel and the part of the lever within the hole in the indexable insert moves towards the rear of the tool and pulls the insert back into the toolholder pocket.

This is a good, rigid clamping system and is suitable for medium and rough turning operations on both steel and cast iron. It also has no loose parts when the insert is being changed. The fact that no overhead clamps are used allows the unrestricted flow of the chips which are produced during machining.

The ISO designation for this system is ‘P’.

Screw Clamping

Figure 52 is an illustration of a screw clamping system. Once more the support pad is permanently fixed into the pocket by a pin which is pressed into the toolholder. In this case this pin has a threaded hole through its centre which the clamping screw utilises. Positive rake inserts with a countersunk or trumpet shaped hole are clamped by this method. The screw hole is offset to the pocket location and so as the screw is tightened it locates on one side of the tapered hole. Positioning is such that as the screw is tightened further and the head goes deeper into the taper the insert is directed to the abutment faces and then locked against them as the screw tightens up. In this way positive rake inserts are pulled down and back at the same time.

It is not a suitable system for roughing operations and so is usually restricted to positive rake inserts. There are no overhead encumbrances with screw type clamping and so chip flow is unrestricted. It has a slight

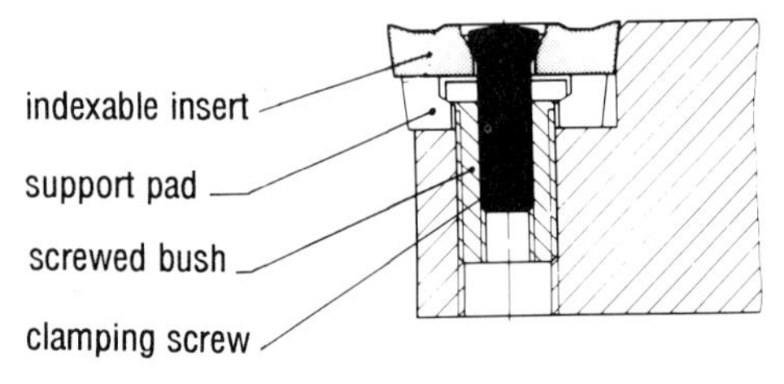

Fig. 52 Screw Clamping of an Indexable Insert

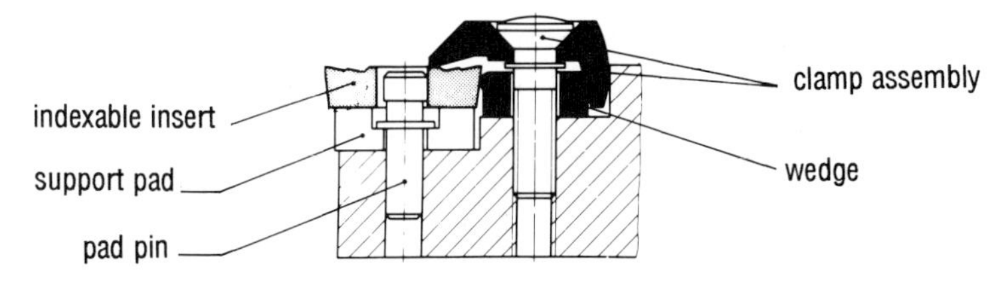

Fig. 53 Wedge Lock Clamping of an Indexable Insert

disadvantage in that the screw must be removed when changing the insert and care must be taken not to lose it in the machine.

The system is designated 'S' in the ISO system.

Wedge Lock Clamping

It is sometimes not possible to fit the lever system into the space available at the head of a toolholder. In such cases a wedge lock clamping system could be the answer. Figure 53 illustrates one example of a wedge lock clamping system. The support pad is held by a fixed pin which has an extension to it and against which the insert is driven by the wedge. In principle a wedge lock system can be used for negative, neutral 7°, and positive rake 11° inserts.

The case illustrated is a neutral 7° indexable insert. As the screw is tightened, the clamp is forced down causing the wedge to be pushed forward. This in turn pushes the insert against the head of the fixed pin. The wedge becomes the abutment face for the insert and the finger of the clamp ensures that the insert does not lift up out of the seating.

The ISO designation for wedge lock clamping is 'M'

The four clamping systems which are illustrated in Figures 50, 51, 52 and 53 are merely examples to illustrate the principle of the type of clamping which is designated in the ISO standard. Each manufacturer will have his own versions of the clamping methods but they will be very similar to the ones described.

There are other specialised clamping systems in existence but they tend to be exclusive to a particular toolholder manufacturer and are not generally used by all suppliers. In Figure 54 four such clamping configurations for ceramic inserts are shown.

With proper maintenance there is no reason why toolholders should not give excellent service. Inserts should be indexed at a predetermined time or piece quantity so that major damage to the toolholder is avoided. Periodic inspection of the clamping elements and support pads is also an important action. Worn or defective parts should be immediately replaced.

Fig. 54 Clamping Methods Ceramic Inserts

An indication of the type of clamping which is most likely to perform well for various turning operations is given in the table below.

	C Overhead clamp	M Wedge lock	S Screw	P Lever lock
Plain O.D. turning	x	x	x	x
Facing in	x	x	x	x
Out facing		x	x	x
Copy turning – in feeding			x	
Copy turning – out feeding		x	x	x
Fine feeds	x	x	x	
Medium feeds	x	x	x	x
Coarse feeds				x

Tool and insert selection

In selecting the tool and insert to be used to carry out a turning operation consideration must be given to the following points.

The Workpiece

The material of which the workpiece is composed and the condition of the workpiece e.g. a forging, casting etc.

The shape and size of the component to be machined and the dimensional tolerances which have to be achieved.

The required surface finish on the machined component.

The Machine which Will Carry out the Turning Operation

The type of machine and the extent of its programming ability.

The tool clamping arrangement, the size of toolholder able to be fitted.

Whether the direction of cutting is right or left hand.

How many tool positions are possible.

The power available from the machine.

The chucking/holding system for the workpiece.

The Cutting Edge Length and Corner Radius of the Insert

Suggestions for the cutting edge length of the indexable insert and corner radius which should be chosen are offered in the table below.

	Insert shape/ cutting edge length						
	C	S	T	R	D	V	Corner radius
Fine feeds 0.8–0.25; small depths of cut 0.3–2.0	06 09	06 09	09 11	06 08 10 12	07 11	11 16	02 04 08
Medium feeds 0.2–0.6; medium depths of cut 2.0–6.0	09 12 16	09 12 15	11 16 22	10 12 16 20	07 11 15	11 16	04 08 12 16
Coarse feeds 0.5–1.5; large depths of cut 5.0–15.0	16 19	15 19	22 27	16 20 25 32			08 12 16 20 24

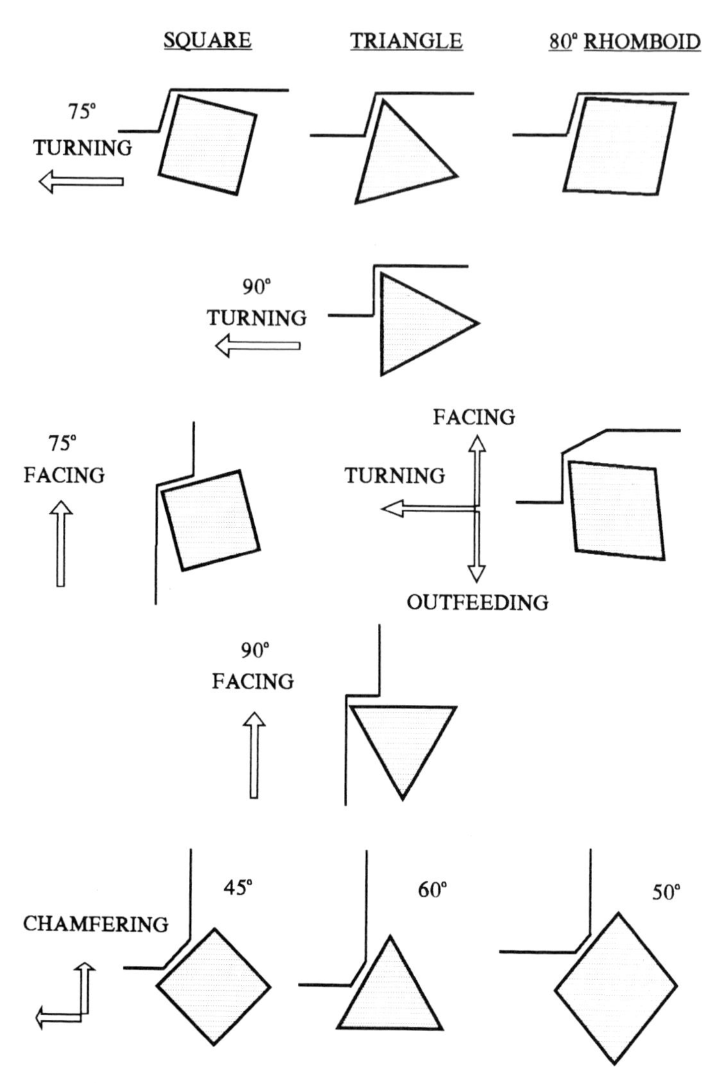

Fig. 55 Possibilities for Turning with Indexable Inserts

For high quality surface finish larger corner radii are advantageous. For less stable and thin walled and slender workpieces smaller nose radii should be employed.

Possible Cutting Profiles Using Indexable Inserts
Examples of how indexable inserts can be used to carry out various turning operations are illustrated in Figure 55.

Fig. 56 Indexable Insert Turning Tools

The introduction of programme controlled lathes increased the versatility of 80° rhomboid inserts which are able to turn, face and outfeed with the same tool.

There are two popular styles of copying insert available which enable undercuts and outfeeding at angles greater than 90° to be made. These are designated 'D' which is diamond shaped with a 55° corner angle and 'V' which again is diamond shaped and has a corner angle of 35°.

A range of indexable insert turning tools (both external and internal) is shown in Figure 56.

1.10 BORING TOOLS

As with turning tools there are three groups as follows:

a) High speed steel tools
b) Brazed and solid hardmetal tools
c) Toolholders with indexable inserts.

High Speed Steel Tools

Round, ground high speed steel tool bits are applied by being clamped in holes appropriately positioned at the end of boring bars. Square sectioned tool bits can be used in the same way but the demand for these is much less. They are included in BS 1296 : Part 4 which has already been described under 5.1.9 – Turning Tools.

Butt welded boring tools are also used for boring and these are included in the British standard BS 1296 : Part 3. Two boring tools are specified in this standard with a reference number 50. One of the tools has a square nose which means the tool can bore and face, the other tool has a round nose. Four preferred sizes are specified.

Brazed Hardmetal Tools

Brazed hardmetal boring tools are supplied in accordance with the old British Hardmetal Association industry standard. As with high speed steel tool bits the most popular styles have round shanks and are clamped in boring bars in the same way as the high speed steel tool bits. The sizes are specified in imperial dimensions. The two smallest diameters in the standard, 3/16″ and 1/4″, are made from solid hardmetal.

A range of square sectioned boring tools is also included which are clamped into boring bars and used in the same way as the round ones.

If the direction of cutting would be termed right hand, which is the usual direction, then the boring tools which are clamped in the boring bar must be left hand.

Tools with Indexable Inserts

As with internal turning tools an ISO designation system for boring tools is in existence. This is ISO 6261 entitled *Boring Bars (Tool Holders with cylindrical Shank) for Indexable Inserts – Designation.*

This designation consists of 9 positions made up of letters and numbers as follows:

Position 1. (Letter) – Type Of Bar

Each type of bar has a single letter to describe it. For example 'S' is the letter used to designate a solid steel boring bar whilst 'A' is used for a steel boring bar which has an internal coolant supply hole passing through it.

A solid steel boring bar which is damped to counteract vibration has the letter 'B'. The same bar but with an internal coolant supply hole is designated 'D'.

'C' covers a bar consisting of a hardmetal shank which has a steel head attached which carries the indexable insert. If the same bar has a coolant supply hole passing through it then it has the letter 'E'.

These are just six of the types of bar which are covered by this position.

Position 2. (two digit number) – Bar Diameter
The diameter of the boring bar is stated in mm. Diameters less than 10 mm are prefixed by a zero e.g. 6 mm diameter = 06.

Position 3. (letter) – Total Length of the Tool
Specific lengths are designated ranging from 80 to 500 mm. The letter 'X' is reserved to indicate that the length is special and does not conform to one of the fixed lengths laid down. The length is defined as the total length of the tool with the insert in position.

Where the lengths coincide with those in the ISO standard for turning tools – ISO 5608 – the same letter is used see 5.1.9 position 8.

Position 4. (letter) – Clamping Method
Once again the letters used correspond to the same methods of clamping which are designated in the turning tool standard which is fully described in 5.1.9.

Position 5. (letter) – Insert Shape
The letters used are the same as those which are described in 3.2 Detail 1. i.e. 'S' for square, 'T' for triangular, 'R' for round etc.

Position 6. (letter) – Style of the Boring Bar
The way in which the tool is presented to the workpiece is designated by a letter. The different possibilities are illustrated in the standard itself.

Position 7. (letter) – 'Clearance Angle' of the Indexable Insert
Although this position is termed clearance angle it does not refer to the actual clearance angle which will exist when the insert is mounted in the toolholder but is 90° minus the angle which the side of the insert makes with its top face. For example an insert with a right angled side has a 'clearance angle' of 0° and is designated by the letter 'N'.

The letters used and the principle of the system is identical with that for designating the indexable insert and is described in 3.2, Detail 2.

Position 8. (letter) – Hand of the Tool
The tool can either be right hand cutting – 'R', left hand cutting – 'L' or able to cut in either direction – 'N'.

Position 9. (two digit number) – Cutting Edge Length
This position defines the size of the insert being used by reference to its

cutting edge length. It is the same two digit number used in chapter 3.2, Detail 5.

Below is an example of this designation system for boring bars:

Example: A20R-SCLCL09

A = A steel boring bar with internal coolant supply
20 = The diameter of the boring bar is 20 mm
R = The total length of the bar is 200 mm
S = The insert is clamped by means of a screw
C = The insert used is an 80°C rhomboid shape
L = The approach angle of the tool is 95°
C = The insert used has a 7° side angle
L = The tool is left hand cutting direction
09 = The designation of cutting edge length of the insert is 9.525 mm

One of the problems which can arise when a boring operation is being carried out is the occurrence of vibration during cutting. This will result in a poorer surface finish, damage to the cutting edge and can also affect the machine on which the work is being done. Special bars are available which deal with the problem of vibration. These anti-vibration bars ensure a better surface finish and less wear of the indexable insert. They also reduce the noise which comes with vibration.

A second problem is the possibility of deflection of the boring bar resulting from the push off force during cutting. The tolerances which can be held on the component being machined are very much dependent on this situation. In order to prevent deflection, as a rule of thumb, the maximum length/diameter ratio for a heat treated steel bar should be 5:1. This means that if a 10 mm diameter steel bar is clamped so that it overhangs then the length of overhang should be a maximum of 50 mm in order to avoid deflection assuming that a reasonable load is applied to the projecting end of the bar. If the material from which the bar is made is changed to hardmetal then the maximum length/diameter ratio increases to 9:1.

All these types of boring bar are covered in the ISO designation standard. They are identified by a letter which is placed at the first position of the designation.

A boring operation requires careful consideration when selecting the tool and indexable insert which will be used to carry out the machining procedure. Because cutting is usually taking place in a confined space chip control is critical. The choice of clamping system used on the boring bar must take account of the type of chips which will be produced. Top clamp-

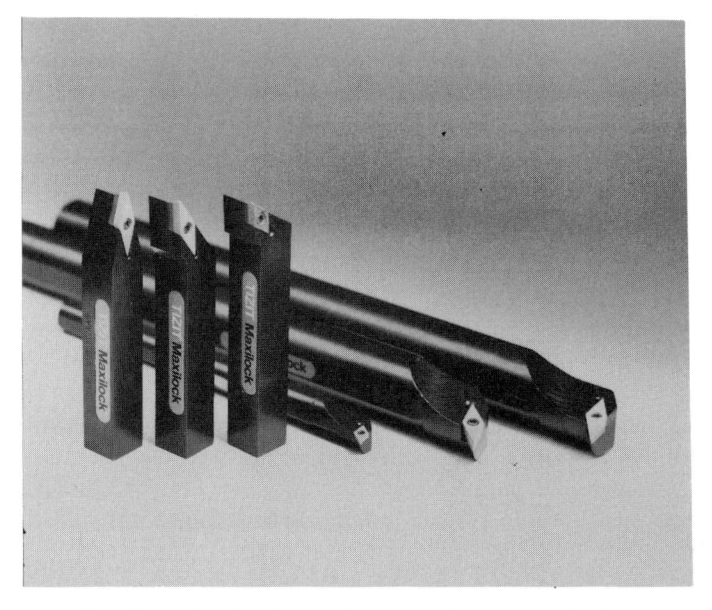

Fig. 57 Indexable Insert Tools for Boring Applications

ing will tend to deter easy dispersion of the swarf. A good flow of coolant can assist in clearing chips and so internal coolant supply should be borne in mind when selecting the boring bar. Neutral or positive rake cutting angles are preferable to negative rakes and the importance of selecting an appropriate chip control groove in the indexable insert cannot be overemphasised. The centre height of the cutting tool should be correctly positioned, if the tool is cutting below centre then the clearance angle is reduced and excessive wear results . Cutting above centre can cause excessive cutting forces. Centre height is also a factor to check if vibration occurs.

The boring bar chosen should have the largest diameter which can be accommodated in the machining operation. The overhang of the bar should be kept as small as possible and the clamping system used to hold the bar should be rigid. The nose radius of the indexable insert should be the smallest which is compatible with the operation being carried out.

Some examples of toolholders for boring are shown in Figure 57.

These are tools used for internal copying. The inserts are fastened to the toolholders by means of a screw clamping system which is designated 'S' in the ISO designation standard.

The photograph also shows external copying toolholders which are similarly clamped by means of a screw through a hole in the centre of the indexable insert.

5.1.11 CARTRIDGES

Perhaps the best description of cartridges is given by the world's largest manufacturer of hard material cutting tools – Sandvik. They use the term 'Build-in' tools. Cartridges are mini toolholders which carry indexable inserts. They can be used for both external and internal work (turning and boring). Their main demand is in multi-cutting edge tools which are usually special to a particular application.

Cartridges for turning and boring are covered by the ISO designation standard ISO 5608 which has already been described in 5.1.9 Turning Tools – Tools With Indexable Inserts. For cartridges it has ten positions and is a letter and number system. The first six positions are identical with the toolholder designation, the remaining four positions are explained below:

Position 7. (letter) – Type of Tool
In this case the letter 'C' is used indicating that the tool is a cartridge.

Position 8. (letter) – Type of Design
Alternative designs are according to ISO 5611

Position 9. (dash or letter) – Tool Length
In this case a dash indicates that the tool length is according to ISO 5611.

Position 10. (two digit number) – Cutting Edge Length
This position defines the size of the insert being used by reference to its cutting edge length. It is the same two digit number used in chapter 3.2, detail 5.

The following example illustrates the use of the cartridge designation system:

Example: SSKNR16CA-12

S = The indexable insert is clamped by means of a screw
S = The cartridge takes a square insert
K = The tool has a 15° approach angle
N = The insert has a right angled side – negative insert
R = Right hand cutting
16 = The cutting height is 16 mm
C = The tool is a cartridge
A = The cartridge design is according to ISO 5611
– = The length of the cartridge is according to ISO 5611
12 = The cutting edge length of the insert is 12.5 mm

The use of cartridges on the main tool body offers several advantages. Both radial and axial adjustments are possible. If a modification is required to the workpiece the adjustment can usually be taken up by the cartridge instead of having to modify the basic tool. Should one indexable insert break then any damage is confined to the build-in tool itself and the expensive special tool is not harmed. One further advantage is that cartridges make it possible to hold relatively small positional tolerances.

Cartridges are used on multi-cutting edge tools as build-in tooling. Typical examples are cases where several different boring positions exist on the main tool and the operations are carried out in one continuous movement. A second situation where build-in tools are used is when it is necessary to distribute the total depth of cut over several cutting edges. This may be either a turning or a boring operation and two or more cartridges may be used.

One special type of cartridge is a boring unit which is capable of very fine adjustment for close tolerance work. These fine boring cartridges can be mounted into boring bars or built into special boring heads.

5.1.12 OTHER SPECIAL TURNING APPLICATIONS

Heavy Duty Turning Tools

For really heavy work, such as turning very large steel forgings and castings, the standard ISO range of indexable insert turning tools would be stretched to perform satisfactorily and may well fail under these extreme conditions. To overcome this the cutting tool manufacturers offer their own designs of heavy duty indexable insert turning tools. The inserts and the clamping systems are not standardised and each manufacturer has arrived at his own design solution.

A range of such tools from one manufacturer is shown in Figure 58. The steel chip illustrated in this photograph gives a good impression of the work undertaken by these tools, it is approximately 25 mm wide and 2 mm thick. Obviously the machine tools used to do this type of work must be sturdy and powerful to remove metal in this way. The hardmetal grades which are employed to machine under these conditions are those designed for the P40 – P50 groups of application.

Bar Peeling

Bar peeling is an operation which removes the outer skin from a round bar and produces an accurate, peeled round bar with a good surface finish.

Fig. 58 Heavy Duty Turning Tools with Indexable inserts

The bars are continuously pulled or pushed through a rotating head which is fitted with either three or four equally spaced hardmetal cutting tools. Bar peeling machines are purpose built and with smaller diameter workpieces can have throughput rates of up to 80 metres per minute.

The process can be divided into three categories:

Light peeling – 6 to 60 mm bar diameter
Medium peeling – 60 to 150 mm bar diameter
Heavy peeling – 150 to 600 mm bar diameter

Small approach angles (15° – 20°) are commonly used which do the primary cutting and then blend into a parallel cutting edge which does the secondary cutting.

Originally brazed tools were used but these are now replaced to a large extent by indexable insert tools. The 'W' style trigon shaped inserts are a popular choice to achieve the required geometry in one insert. For heavy work cartridges can be employed which use two or more inserts. The first inserts do the roughing work, dealing with scale, the uneven surface and any cracks. The final insert has the parallel form which sizes the bar and promotes a good surface finish. In order to achieve a surface finish which is free from steps the feed rate must be smaller than the length of the secondary cutting edge, i.e. the secondary cutting edge must overlap the feed. For even better surface finish burnishing rolls can be applied as a final operation.

By using a shallow approach angle the primary cutting edge is protected against irregular shaped bars and scale. It also reduces the chip thickness and leads to higher tool life. The edge condition of the inserts is an important factor especially with heavier work.

The cutting materials which are used and the cutting data depend on the workpiece material to be machined and the diameter of the bar itself. The power available and the rigidity of the machine are other influencing factors.

For ferritic steels, hardmetals which fit the P10 – P30 application groups are normally used. For stainless steels a K30 application group hardmetal is popular. Coated hardmetals are becoming firmly established.

The nimonic series of heat resisting alloys are extruded into bars at high temperature using glass as a die lubricant. On cooling down these bars have an oxide skin and also have glass trapped in their surfaces. The cutting material used to peel these bars must be capable of coping with the fluctuating loads from the uneven surface and must resist the abrasive wear both from the oxide skin and the glass and finally must cut this most difficult to machine workpiece material. Before coatings were introduced the choice was a compromise and a K40 application group grade was used. Today tough coated grades of hardmetal give excellent results.

With more difficult work the cut can be broken up into stages by using a tool consisting of more than one indexable insert. Round inserts can be employed for the more arduous work and Figure 59 shows a tool with two round inserts and a triangular insert with a relieved corner. The depth of cut is divided between the two round inserts and the sizing and finishing is performed by the triangular insert. A choice of different hardmetal grades can be selected, tougher for the round inserts and more wear resisting for the triangles.

Depths of cut used in bar peeling depend on the size of the inserts used and also on their shape but in the limit they can go up to 10 mm. The feed of the bar through the machine ranges from 3 to 50 metres per

Fig. 59 Special Indexable Insert Tools for Bar Peeling

minute and with smaller bars can go up to 80 metres per minute. Cutting speeds are from 40 to 200 metres per minute depending on the workpiece material and the bar peeling machine being used.

5.1.13 GENERAL POINTS AND ADVICE

1. **Never** stop the machine whilst the tool is in cut.
2. Remove the keen edge from a reground brazed hardmetal tool by one or two gentle strokes with a diamond or boron carbide hand lap.
3. Rigidity of the machine and the workpiece holding system is a critical factor for successful results when machining with hard cutting materials.
4. Ensure that the tip seat of an indexable insert toolholder or milling cutter body is clean and free from debris before replacing the insert or breakage may occur.
5. Choose bigger indexable inserts for heavier, roughing cuts and smaller inserts for lighter cuts.

6. Uncoated hardmetal grades normally produce a better surface finish than coated hardmetal grades.
7. Excessive tool overhang leads to vibration. Overhang should be the smallest possible to maximise rigidity.
8. Check that the tool is cutting on centre. Centre height is important with the lighter finishing cuts and especially with parting and grooving.
9. When using uncoated hardmetal always use the plain Co-WC grades unless machining ferritic steels (they will form a crater). The Co-TiC-WC grades, the so called Steel Cutting Grades are often believed to be the tougher group of grades because they are used to machine ferritic steels but this is not so, the Co-WC grades are the toughest!
10. The maximum recommended depth of cut with round indexable inserts is one quarter of the diameter of the insert.
11. Use ceramics only where hardmetals would perform without problems – but increase the cutting speed.
12. Do not use ceramics to solve machining problems – use them for cost reduction.
13. When machining with ceramics use round or square indexable inserts with large nose radii wherever possible to obtain maximum strength.
14. Pre-chamfering of the workpiece at the point of entry of the cut can be helpful, particularly in the case of ceramics. A pre-chamfer at the end of the cut may also assist in preventing breaking off from the workpiece as the cut is finishing.
15. Care should be taken when hand grinding high speed steel, stellite and hardmetal tooling. It is advisable to remove from the vicinity any containers which have water in them. Quenching the tools in water to cool them down when they have become hot during hand grinding is fatal – they will crack!

5.2 PARTING AND GROOVING

Cutting tools for parting and grooving fall principally into three groups and these are:

a) Solid high speed steel or solid hardmetal parting and grooving blades.

b) Butt welded high speed steel and parting/grooving tools and blades which are tipped with hardmetal – in this case the hardmetal is brazed onto the tool or blade.
c) Parting/grooving tools and blades where the cutting material is an insert which is clamped to the tool by one of a variety of holding systems.

These are the most common tools and materials involved but there are special cases where other cutting materials are used such as grooving aluminium alloy pistons with PCD. The surface finish and repeated accuracy of the groove, all at very high cutting speeds, make PCD the ideal choice.

High Speed Steel and Solid Hardmetal Blades

High speed steel blades are offered in both metric and imperial dimensions. Their widths range from 2.5 to 6 mm. Lengths range from 90 to 200 mm according to the blade width and type. They are clamped in purpose built holders and when the cutting edge is worn the blade is removed for regrinding.

Solid hardmetal blades are used for producing the grooves which hold the piston rings in aluminium pistons.

Butt-Welded HSS and Brazed Hardmetal Tools

Butt-welded high speed steel tools are supplied to the British standard BS 1296 which is fully described in 5.1.9 Turning Tools – High Speed Steel Tools. These are available in a variety of parting widths and also either left or right hand cutting.

Similar tools are produced which have hardmetal tips brazed on to provide the cutting edge. They conform to either the British industry standard for brazed tools or to the ISO standard – ISO 243 Tool No. 7.

The main advantage with both the butt-welded high speed steel and the brazed hardmetal parting tools is that any desired geometry can be ground onto the tool. Side clearance is critical to the performance of the tool and can easily be adjusted with these tools by grinding but is a much more difficult task with a solid blade. Counter to this is the much shallower depth of penetration possible compared with solid blades.

Clamped Insert Tooling

In the past the major deterrent to the use of clamped insert parting and grooving tools was the difficulty of designing a satisfactory clamping system. With penetration depths of more than ca. 15 mm the parting insert must be held from above and below and at the same time to be viable in the market the total width of the blade and clamp should not exceed 3 mm. It is not a simple task to design a clamping system with such capability. Credit must be given to ISCAR for introducing the most successful development in the clamping of parting and grooving inserts. In the trade this is known as the 'Self Grip' system and many variations are now offered by major cutting tool manufacturers. The principle of the self grip system is shown in Figure 60.

A blade which can be no thicker than the insert it is designed to carry, say 3 mm, has a slot cut in each end (this makes it reversible should one end become damaged). The insert is pushed into this slot causing it to spring open slightly and grip the insert. The insert becomes a wedge which jams into the slot. The angles of the wedge shaped slot have been precisely calculated to produce the maximum grip when the insert is

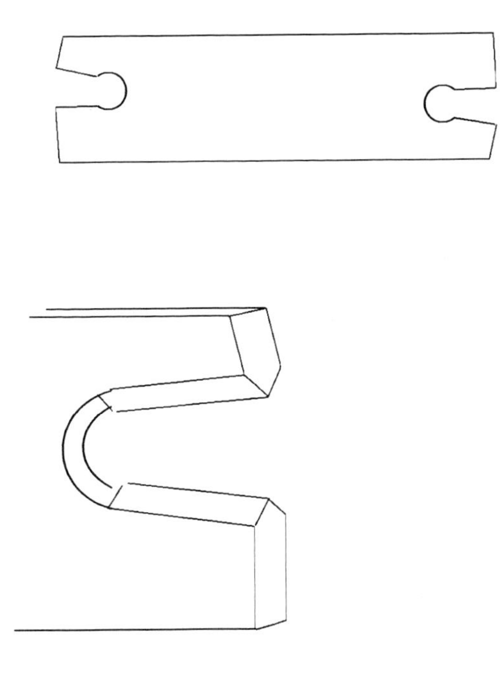

Fig. 60 Self Gripping System for Parting and Grooving Tools

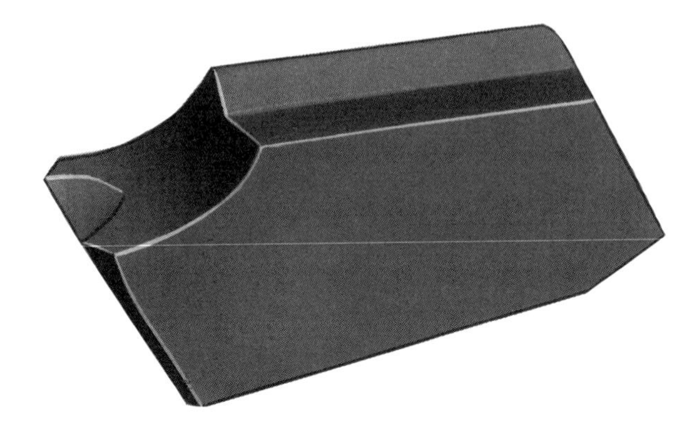

Fig. 61 Self Gripping Indexable Insert for Parting

pushed home. The steel and its heat treatment are chosen so that the appropriate strength and wear resistance are provided. Most designs have some form of stop incorporated so that the insert cannot be driven in too far. Special keys are used to facilitate ejection of the insert.

The upper and lower faces of the slot have 'V' projections running along their length and the inserts have matching 'V' shapes recessed into their top and bottom faces. A photograph of a typical hardmetal insert is shown in Figure 61. The 'V' formed locations ensure alignment of the insert both horizontally and vertically.

The blades are held in special holders allowing blade changing and setting to be carried out easily. Figure 62 shows examples of self grip type tools and inserts.

Where penetration depths are smaller two types of insert holding can be used. The first is by means of an overhead clamp and in this case the same 'V' shaped projection described above may be used on the clamp and on the insert seating so that the inserts used in the self grip blades are interchangeable with the overhead clamp type tools. Other configurations of clamp and seating using appropriately designed inserts also exist.

The second system uses thin, square or triangular blades which have a hole through their centre. These are mounted vertically onto a holder and are clamped by means of a screw or a 'pull back' device. High speed steel, coated high speed steel or hardmetal blades are available for this

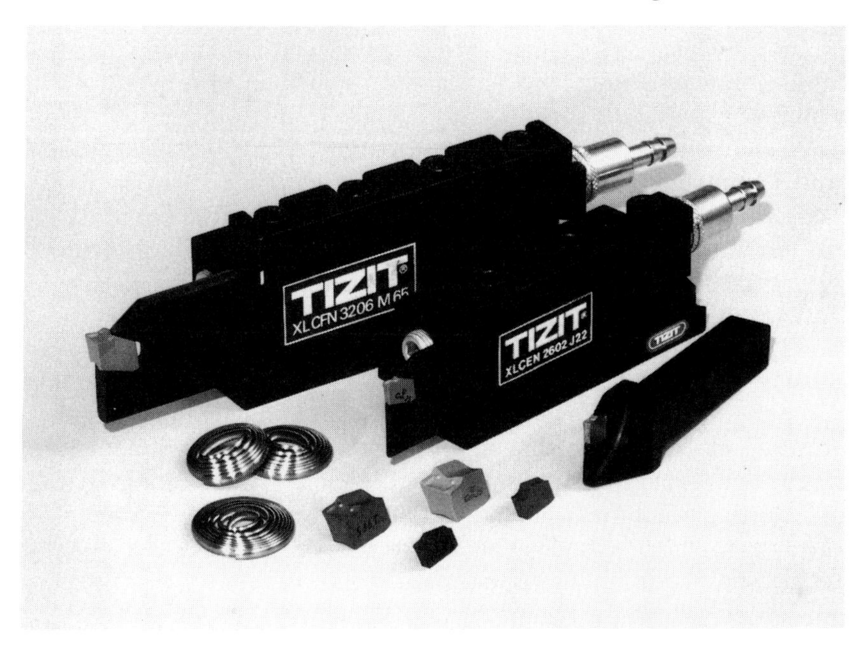

Fig. 62 Self Gripping Type Tools for Parting and Grooving

type of parting and grooving tool. They are most suited to parting off thinner walled tubes and for producing shallow groves such as circlip or O-ring grooves.

General Points on Parting

When parting off from bar, if the rotational speed of the workpiece remains constant, i.e. the spindle speed is constant, then the cutting speed reduces as the cutting edge approaches the centre of the bar and at the centre itself the cutting speed is zero. This situation is generally acceptable for high speed steel as high positive rake angles can be used. However, with hardmetals this reduced cutting speed needs the employment of the tougher grades of hardmetal to overcome the possibility of breakage and a constant cutting speed should be employed wherever possible. Modern machine tools are capable of variable speed operation and this ensures that hardmetals will perform satisfactorily. With parting and with grooving the chip is being formed within a very narrow slot and it is essential that it is transported away. To this end cutting edge geometry is important when using hardmetal and this is particularly so with large penetration depths. The ideal geometry will cause the chip to have a slightly concave cross section – incurved – which effectively reduces its width and allows it to clear the sides of the groove which is being machined so that it can get away without fouling the groove. The best chip form is probably a flat spiral coil, in other words like a clock spring; examples of this type of chip are included in Figure 60. The use of cutting fluid (coolant) is always recommended and a copious supply is advised.

The centre height of the cutting tool is very important when parting. A tolerance on the centre height of +/- 0.1 mm is essential when parting off from bar stock and is strongly recommended for parting off tubes and for grooving. The overhang of the cutting tool should also be kept to a minimum to maximise rigidity and prevent the setting up of vibration.

Cutting speeds with hardmetal tooling are usually of the order of half those which would be used for turning the same workpiece material. A good starting point for the cutting speed is 100 metres per minute.

The range of feeds used for parting starts from as low as 0.05 mm per revolution and goes up to 0.5 mm per revolution with the main area of application lying in the region 0.1 to 0.2 mm / rev. Larger feeds can be employed with larger blade widths and with smaller depths of penetration and the smaller feeds are used in the opposite situations.

In order to save workpiece material when parting off, narrow cutting edges are used. The stability of the tool system is a limiting factor to how narrow one can go. In addition the side clearance angles required on the tool can also limit the width of the cutting edge. The following list is a guide to the maximum diameter of workpiece which should be parted off for a given cutting edge:

Width of tool (cc)	Maximum diameter of cutting edge (cc)
2	50
3	70
4	80
5	100
6	120
8	150
9	150

In parting off there is a tendency for the part which is being separated to break off before the cut comes finally to the centre of the bar and this leaves a pip on either the bar or the piece which has been cut off or on both. This is especially so if the front face of the cutting edge is parallel to the axis of the bar. In the case of tubes a ring or burr is formed. By applying an angle of 4° to the front edge of the cutting tool no pip is left on the part adjacent to the leading corner of the tool. If one stands in front of the machine with the workpiece on the left then a right hand tool has the leading corner on the right and leaves the pip on the bar and not on the piece parted off. With a tube the same thing applies, the burr or ring is left on the tube and can be machined off before the next piece is removed. Reducing the feed rate is also a help if difficulties still persist.

It is not good practice with hardmetal to machine past the centre as this can lead to breakage of the cutting edge.

Coated grades of hardmetal are now very popular with clamped parting tools. They give higher tool life and tend to reduce the formation of a built up edge. For steels, the tougher P30 -P40 ISO application group grades are the most popular. Suggestions for the feeds which should be used according to the width of the tool are given below:

Width of tool (mm)	Feed range (mm/rev)
2	0.03–0.12
2.5	0.03–0.18
3	0.05–0.25
4	0.1 –0.3
5	0.1 –0.35
6	0.1 –0.4

For feeds of 0.2 mm/rev. and below a positive cutting geometry together with a sharp edge will give an advantage. This is especially so with thin walled tubes and slender components.

General Points on Grooving

Popular grooves needed are those for circlips and O-rings. Face grooving is also another operation required to be carried out. Many of the points which have already made for parting operations are valid for grooving but the profile of the cutting edge will be according to the form required at the base of the groove. With grooving the feed should be related to the surface finish demanded. Finer feeds will produce a better surface finish and values as low as 0.03 mm / rev. are employed. With hardmetal it is not desirable to dwell the tool at the base of the groove to try to improve the surface finish.

The following lists give recommendations for the feeds which should be applied according to the width of the grooving tool. The first list covers normal grooving (internal or external) and the second list applies to face grooving.

Normal Grooving

Width of tool (mm)	Feed range (mm/rev)
2	0.03–0.12
2.5	0.03–0.15
3	0.03–0.15
4	0.0.5–0.2
5	0.0.5–0.2
6	0.05–0.2

Face Grooving

Width of tool (mm)	Feed range (mm/rev)
2	0.03–0.05
2.5	0.03–0.08
3	0.03–0.1
4	0.05–0.1
5	0.05–0.1

Because feeds are usually lighter than with parting and penetration depths are generally smaller then where hardmetal grades are used they can be more wear resistant. For steel machining the P20 ISO application group grades are the popular choice and with non-ferrous metals the K10 – K20 ISO group grades of hardmetal are mostly used.

5.3 THREADING

The production of threads by rolling techniques involves the deformation of the blank being threaded as opposed to cutting it. This book deals with cutting tools only which in the case of threading means screw cutting (turning) and thread milling, thus thread rolling dies are not included in this chapter.

One of the most popular tools for threading consisted of a solid piece of hardmetal with a dove tail cross section which was used to clamp the threading tool into the machine. These solid pieces of cutting material have grooves, which are the teeth, ground along their length so that when they are mounted tangentially to the workpiece the teeth become the thread form required to be cut. As the cutting edges wear, the top of the tool is reground to just below the depth of the clearance face wear and then it is remounted at the correct cutting height in the machine. The tool has to be inclined towards the workpiece so that clearance can be achieved. Usually three or four teeth are ground in with the leading tooth being flat topped and shorter than the rest which gradually increase in length up to the last tooth which then has the full thread profile. Thus each tooth takes a small cut and the final tooth does not have an excessive amount of work to do. This multi toothed system is important for larger threads and particularly so for the oil industry. However, for smaller threads single tooth cutting tools are viable and in this case small hardmetal pieces are brazed onto steel bodies which have the same dove tail cross section for holding purposes. Regrinding one small tooth is a comparatively simple matter and this is a much more economical system for small threads than using a solid piece of hardmetal.

5.3.1 CLAMPED TOOLING

Indexable insert technology is now strongly established in threading operations. This area is almost entirely dominated by hardmetal as a cutting material and by using indexable inserts full advantage has been taken of the possibilities of coatings for improving performance and productivity.

Figure 63 shows a collection of clamped threading tools and indexable inserts. One of the favoured shapes of indexable inserts for threading is a triangle with a thread profile ground onto each of the three corners. Screw clamping or top clamping are the two most popular methods of holding.

The types of insert which are available can be divided into two broad groups:

Single toothed inserts
Multi-toothed inserts

Single toothed inserts are further divided into two types – partial profile and full profile inserts.

Partial Profile Inserts

With partial profile inserts the process of forming the thread does not include the finishing of the outside diameter of an external thread or the

Fig. 63 Threading Tools with Hardmetal Indexable Inserts

finishing of the inside diameter of an internal thread. This means the thread is not directly calibrated during the thread cutting operation.

Advantages:

- An insert may be used to produce threads with the same tooth angle and having a pitch within a specified range for that insert.
- A smaller stock range is needed.

Disadvantages:

- The radius at the top of the tooth profile is the radius for the smallest pitch within the specified range for that insert.
- It is necessary to carry out an additional turning operation to finish the crest of the thread.
- The thread depth and the radius at the root of the thread are not exactly to standard.

Full Profile Inserts

A full profile insert will form a complete thread profile including the crest diameter.

Advantages:

- The thread profile is exactly according to standard.
- The outside diameter is already machined and there are no burrs.
- In the case of components which have both an external and an internal thread then these are concentric.

Disadvantages:

- An insert can only be used for the one pitch.
- A larger stock range is needed.

A thread cannot be produced in one pass. With both single and multi-toothed inserts it always requires several passes to complete the thread. With single toothed inserts the number of passes required is in the range 10 to 15.

Multi-Toothed Inserts

The design of these inserts is such that the subsequent tooth will cut deeper than the one preceding it. The last tooth is the only one with the full form.

Advantages:

- The thread profile is exactly according to standard.
- The number of passes required to produce the thread is less than with single toothed inserts.
- Tool life is generally longer as there is less work load per tooth.
- Productivity is higher.

Disadvantages:

- The undercut at the end of the thread must be large enough to allow for the entire row of teeth on the insert to run out.
- Cutting forces are high giving a risk of vibration. These inserts should be used with stable workpieces also with good machine and tool rigidity.
- They can only be used as full profile inserts.
- Problems can occur with space availability when machining in small areas.

The rake and the clearance angles on the threading inserts must suit the helix angle being machined. In order to achieve the best result the clearance angle on each side of the tooth should be equal when cutting. This is arrived at by inclining the insert to the same angle as the helix angle of the thread. By supplying a support pad with the corresponding inclination already ground in the correct attitude in the tool for the insert is achieved. If the inclination is not at the correct helix angle one of the flanks of the tooth will wear too quickly and the insert life will be short. A differently ground support pad is needed for threads with a different pitch.

With threading tools the support pad has three functions which are:

To allow changes in the helix angle without changing the insert.
To support the threading insert.
To protect the insert seat in the tool.

Internal and external thread forms have differing thread depths and differing radii at the point of the thread and so separate inserts are required for each operation. A second feature is that inserts for internal work have larger clearance angles at the point than those for external machining e.g. 15° instead of 10°.

To machine left or right hand threads the total system must be correctly put together:

RH Threads = RH Toolholder, RH Insert, RH Support Pad and RH Turning.
LH Threads = LH Toolholder, LH Insert, LH Support Pad and LH Turning.

When cutting with single toothed inserts the passes may be stepped radially or stepped radially and on the flank of the tool. With multi-toothed inserts the stepping must be radial.

Radial infeed is the most commonly used method of making a pass and on many machine tools it is the only possible way. It is suitable for fine pitches and cases where no chip flow problems exist i.e. it is good with short chipping materials. It should be the first choice for threading work hardening materials such as austenitic stainless steel. However, when used for coarse pitches there is a risk of vibration and poor chip control.

When the passes are made by stepping on the flank of the tool chip control becomes similar to that when turning. Cutting forces are reduced and therefore wear is less. There is a reduced risk of vibration and the flank of the thread has a better surface finish. This method of stepping is suitable for coarse threads and especially for internal threading when problems of vibration or chip evacuation occur.

Hardmetals are the popular cutting materials in the ISO application groups P01 to P30 and K10 to K20. Coated hardmetals play an important role on some problem materials. In this case PVD coatings with sharper edges are good. Cermets are also being applied in this area.

Cutting speeds are similar to those used in grooving and radial feeds are the same as those used for very light grooving.

5.3.2 THREAD MILLING

Although it is strictly a milling operation, i.e. it is carried out by using a rotating cutting tool similar to an end mill, it is a unique process and it is therefore logical to cover it in this section of the book.

The thread milling tool employs one cutting insert which is mounted onto the end of the cylindrical cutter body so that the teeth on the insert project from the circumference of the cutter and lie parallel to its axis. The insert is usually clamped by means of a screw and an illustration can be seen on the right hand side of Figure 63. The cutter rotates on its own axis at high speed and then whilst it makes one revolution (360°) around the workpiece it moves vertically one pitch length at the same time.

Thread milling cutters are used to mill both external and internal threads. They can be used to thread blind holes without needing to have a thread relief groove. They are particularly suitable for very large workpieces and for non-rotational, non-symmetrical parts. This method of producing threads is good in the case of large bore diameters and is also suitable for interrupted cutting.

The advantages of thread milling are:

- Machining time is short.
- There is no upper limit on the diameter which can be threaded.
- One tool will produce both right and left hand threads.
- The tool takes interchangeable inserts with standard profiles.
- Short chips are formed, so no swarf problems.
- A high surface finish is achieved.
- It is possible to machine hard materials.
- In the case of large diameters the power requirement is lower than with other threading methods.

The usual cutting materials adopted are the range of hardmetals in the ISO application groups P10 -P30 and K10 -K30 and coated grades are the most favoured ones.

5.4 MILLING

All the machining operations which have been discussed in previous chapters have involved a workpiece which is revolving on a fixed axis and a cutting tool which is brought into contact with this workpiece. The tool is then moved in the required direction to machine the workpiece.

With the standard method of milling the tool rotates on its own axis, in a fixed position, and the workpiece is brought into contact with this rotating tool. The workpiece is then moved in the required direction to carry out the machining process.

This difference in the way of machining brings about a difference in the chip formation when cutting. In turning operations, except for profiling, the depth of cut and the feed generally remain constant and so the chip has a constant cross section. With long chipping workpiece materials, unless there is an interruption in the cut, the chip is continuously flowing over the cutting edge throughout the time the tool is carrying out the machining pass (remember that chip control grooves are positioned behind the cutting edge). With milling tools each cutting edge is in and out of cut for each revolution of the tool and this interrupted cutting action causes thermal cycling of the cutting edge by heating up each time it is in contact with the workpiece and cooling down each time it leaves the cut. The cutting materials used for milling tools must therefore be capable of withstanding both the repeated mechanical and thermal shock which results from a milling operation.

As already stated the chip formation when milling is different from that when turning. There is no constant chip thickness when milling. Dependent on the rotation of the milling tool with respect to the traverse of the workpiece the chip will either start with nil thickness and increase to its maximum thickness as the cutting edge reaches the end of the cut or vice versa. This point is explained more fully later in this chapter.

For many years high speed steel was the most used cutting material for milling. It was not until the establishing of hardmetal indexable insert milling cutters that the share of high speed steel in the milling tool market began to decline. Even today 40% of the market is held by high speed steel.

Milling tools can be divided into two broad groups. These groups are defined by the way in which the milling tool is presented to the machine tool spindle for fastening. The first group consists of tools which have a shank which fits into a device in the spindle and is then fastened by an appropriate method. They are called Integral Shank Tools. The second group is the full spectrum of milling cutters which are attached to the spindle by means of an adaptor or arbor.

5.4.1 INTEGRAL SHANK TOOLS

The way in which a milling tool is connected to the spindle of the milling machine limits its diameter. It is not practical to use shanks much above 32 mm. diameter on Shank Tools.

There are three basic types of shank used and these relate to the means of fastening the cutter to the spindle.

a) Straight Shank
The machine spindle has a collet chuck fitted into the spindle into which the shank is mounted and the collet is then tightened.

b) Flatted Shank
The shank has a flat machined on its side. The machine spindle has an arbor fitted into which the milling cutter is mounted. Either one or two grub screws are then tightened onto the flat to fasten the milling tool securely into the arbor. This flat is sometimes machined so that its face is inclined to the axis of the shank and this is referred to as the 'Whistle Notch' method of fastening.

c) Screwed Shank
The milling tool has a plain cylindrical shank which is threaded at the end. This screw thread is then used to fasten the tool into the arbor or adaptor fitted into the spindle of the milling machine.

The screwed shank method of fastening is the most popular as it is probably the most secure and also the most reliable. The strength of the fastening method is important particularly with milling tools with spiral flutes. In this case the spiral flutes can act in the same way as a thread and can try to pull the tool out of the spindle.

It should be noted that shank tools also exist which have a Morse taper as their means of connecting to the spindle.

There are two basic types of shank cutters and these are known as

Slot Drills and End Mills. Slot Drills usually have two or three 30° spiral flutes which are ground to produce a cutting edge all along their length. The end of the tool is also ground so that the tool will cut like a drill as it is plunged into the workpiece. In the case of three fluted slot drills one of the cutting teeth at the end of the drill is ground so that it cuts over centre i.e. that cutting edge is longer than the other two. Slot drills can therefore plunge into the workpiece and can also cut on their side. They can be used to produce enclosed pockets in a workpiece. Figure 64 illustrates some slot drills.

End Mills cannot plunge directly into the workpiece. They must enter from the side and cut on their periphery. Smaller diameter end mills have two or three spiral flutes and the cutting edges run along them. Their helix angle is normally 30°. Medium diameter cutters have four or six flutes and the largest sizes have eight flutes. A large end mill with eight flutes is shown in Figure 65. This photograph also shows the screwed end on the shank which is used to secure the cutter in the machine spindle.

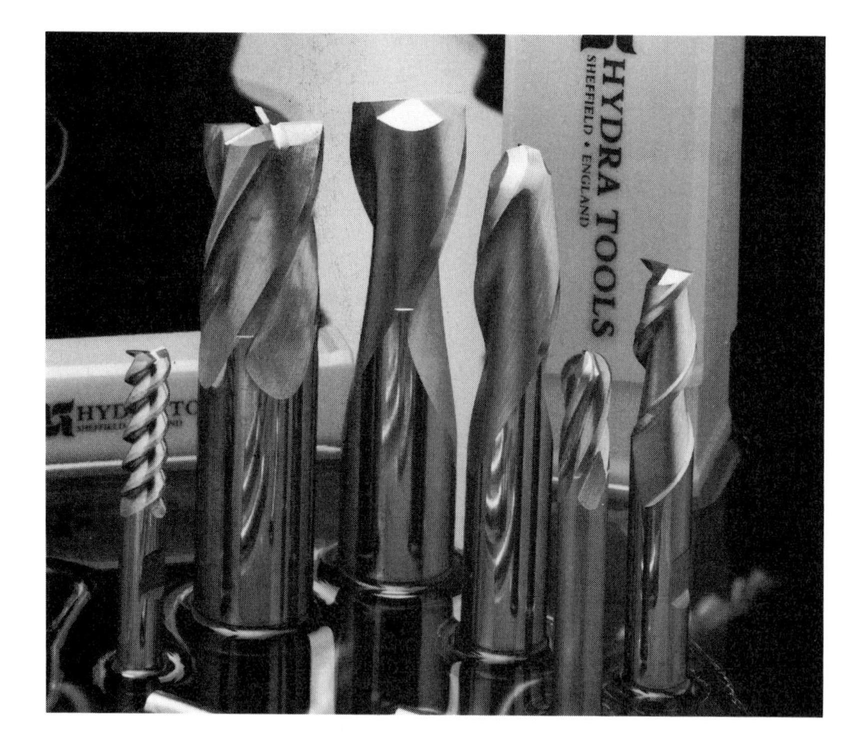

Fig. 64 High Speed Steel Slot Drills and End Mills

Fig. 65 8 Fluted High Speed End Mill

For roughing operations end mills with ribbed spiral teeth are used and examples of these are shown in Figure 66. This photograph shows solid high speed steel roughing end mills with flatted shanks. With roughing cutters the helix angle tends to be less than with the standard end mills and is of the order of 25°.

The majority of slot drills and end mills are made from high speed steel. Applying a TiN coating to these high speed steel milling tools is becoming popular and enables higher metal removal rates to be achieved. Hardmetal is the other main cutting material used for shank cutters and these can be made from solid hardmetal for the smaller diameter tools or indexable inserts tools for the medium and larger diameters. Tools with brazed hardmetal cutting edges are also available.

Relevant standards for shank cutters are BS 122 pt. 4 and ISO 1641/1.

Fig. 66 High Speed Steel End Mills for Roughing Applications

5.4.2 ADAPTOR AND ARBOR MOUNTED TOOLS

Larger diameter cutters are mounted by means of adaptors or by using arbors. The adaptors have tapers which conform to a variety of standards. Some tools have the taper as an integral part of the cutter and this is especially so in the case of brazed hardmetal helical cutters – see Figure 67. In this photograph the two larger tools which are standing up are made from solid steel with hardmetal tips brazed on. The remaining tools are brazed hardmetal tools with cylindrical shanks.

Fig. 67 Brazed Hardmetal Helical Milling Cutters

It is in this field of larger cutters that hardmetal indexable inserts have made the greatest impact in milling and take first place in popularity over high speed steel tools. A typical indexable insert cutter is shown in Figure 68. In this case the inserts are fastened by means of a screw and are mounted with positive axial rake. The tool is a square shoulder face milling cutter. The most popular tools in this category are face mills which have insert pockets directly machined into the cutter body. These are offered in a range of diameters around 40 to 140 mm diameter and with approach angles from 45° to 90°. Square,

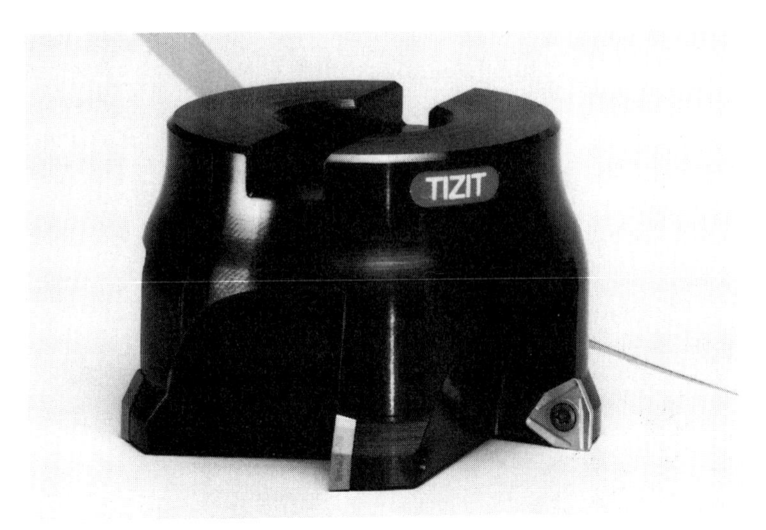

Fig. 68 Indexable Insert Square Shoulder Face Milling Cutters

rectangular, triangular and round inserts are involved and rake angles vary from high positive to negative.

Modular cutters are available which use a variety of cartridges and indexable insert styles. With one cutter body, by selecting the appropriate cartridge and insert combination, the correct cutting geometry for the machining task and for the workpiece material is realised. These modular cutter bodies are offered in a diameter range of 80 to 400 mm.

A range of hardmetal milling tools is shown in Figure 69. Most of these are cutters which incorporate indexable inserts. A modular milling cutter can be seen at the left hand side of the photograph.

5.4.3 SPECIAL MILLING CUTTERS

With mass production operations indexable inserts are an ideal starting point for the design of special milling cutters and there are many examples of these in use in industry. Sometimes cutters are ganged together which is a common method of producing a profile.

There are two special milling tools which are interesting to highlight. The first example is the range used for thread milling which has been discussed in chapter 5.3. The second example is that of the special cutters used for crankshaft milling.

The milling cutters which machine crankshafts pass down the webs

Fig. 69 Examples of Hardmetal Milling Cutters

and machine the cheeks of the crankshaft and then finally mill the pin. The crankshaft is rotating whilst the cutter is being fed forward towards the centre of the component. External crankshaft milling cutters cut on both sides and along their periphery and can be of the order of one metre in diameter. Left and right hand radius inserts are used on each side to machine the cheeks and form the radii which blend the cheeks to the pin. Square inserts are then set around the periphery which cut along their full side to form the pin. More recently cutters are used which are a ring with inserts mounted on their inner circumference. The milling technique is known as crankshaft whirling and in this case the crankshaft passes through the central hole in the cutter which is then rotated around the crankshaft and milling takes place as before. A photograph of a crankshaft whirling cutter is shown in Figure 70. The advantage of this technique is that a smaller number of inserts is required to do the operation than is needed for the external cutter.

Coated hardmetals are used for this milling task which falls in the P30 ISO application group range. Apart from the interrupted cutting of the milling operation the cutting material has to cope with the forged skin of the crankshaft.

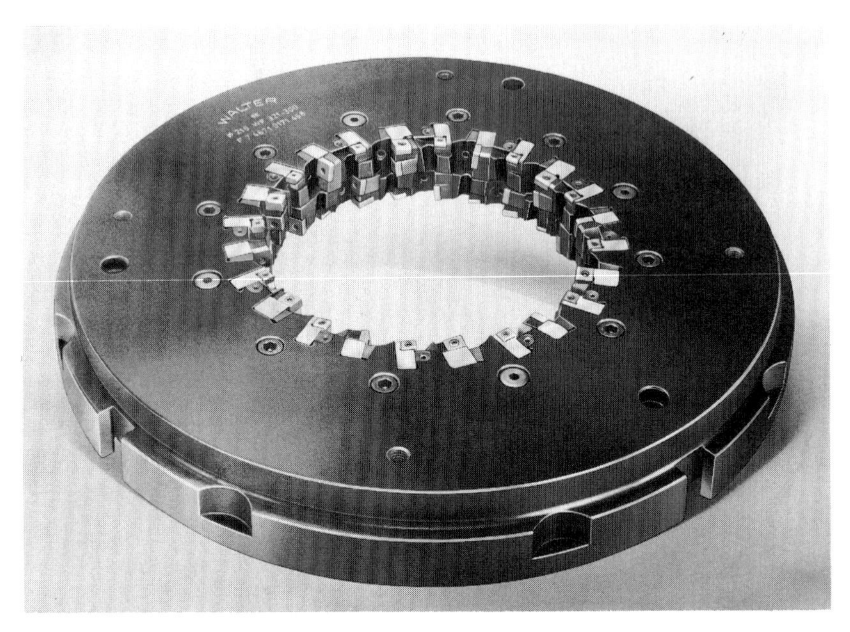

Fig. 70 Crankshaft Whirling Milling Cutter

5.4.4 FACTORS INVOLVED IN MILLING

When a milling tool is cutting on its periphery there are two possible ways that the chips will be produced and these depend on the movement of the teeth on the cutter in relation to the movement of the workpiece.

If the tooth is moving in the opposite direction to the workpiece then the initial contact of the tooth with the workpiece will result in a chip formation such that the chip starts by being very thin and increases to a maximum at the end of the cut. When the teeth come into contact with the workpiece in this way it will result in a squeezing action before the teeth start to bite into the workpiece. This type of milling is known as Conventional Milling or Up Milling.

If, as the tooth enters the cut, it is moving in the same direction as the workpiece then it will take the largest bite at the beginning of the cut and the chip will taper to nothing as the tooth reaches the end of the cut. This is known as Climb Milling or Down Milling. Figure 71 illustrates these two ways of milling.

In general climb milling is to be preferred and this is particularly so

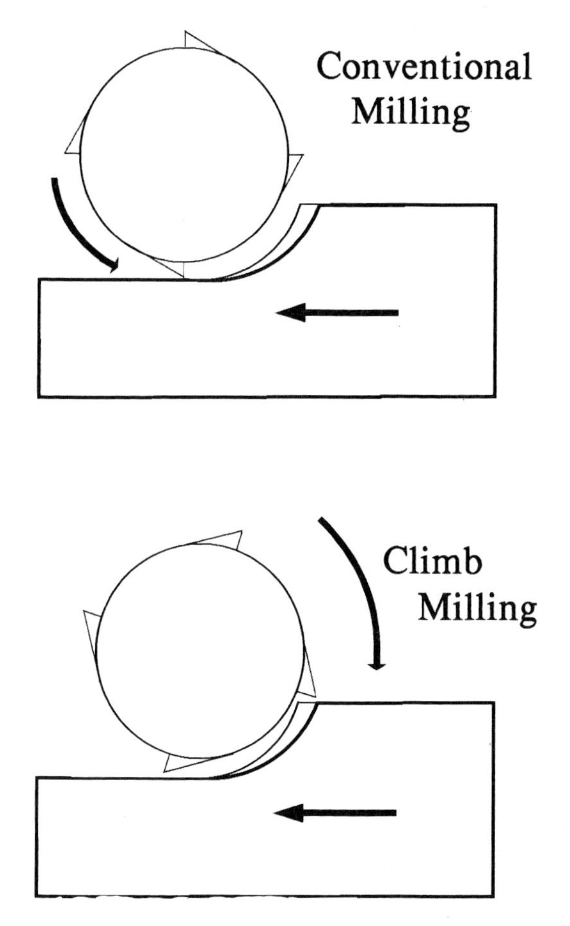

Fig. 71 Diagram Showing Conventional and Climb Milling

when hardmetals are used as the cutting material. With climb milling the main advantage is that the cutting forces are lower and therefore less power is required to remove a specified volume of workpiece material. A second advantage is that the cutting action tends to push the workpiece down and into the clamping fixtures which hold the workpiece and this helps to maintain rigidity.

Conventional milling has the advantage with thin wall section workpieces but does produce higher cutting forces and gives poorer tool life.

Three very important factors in milling which can be problem areas and which need careful consideration are Power, Rigidity and Workpiece Holding. The power which is available from the machine tool limits the way in which the milling cutter can be employed. An old rule of thumb for milling is that the removal of 1 cubic inch of metal per minute requires 1 horse power. Taking 1 horse power as 750 W and rounding off 1 cubic inch to 16.4 cm^3 then this rule of thumb equates to the statement that removing 22 cm^3 of metal per minute requires 1 kW. Obviously this is only a very broad guide.

Using hardmetal as the cutting material, mean values of metal removal rates quoted by one milling cutter manufacturer are:

Aluminium and Aluminium alloys	– 80 $cm^3\ kW^{-1}\ min^{-1}$
Brass, Bronze, Copper	– 45
Grey Cast Iron	– 42
Ductile Cast Iron	– 31
Free Machining and low C Steels	– 24
Alloy and medium to high C Steels	– 22
Tool Steels, soft condition	– 17
Stainless Steels	– 17
Titanium and Titanium alloys	– 14

These values apply to milling with neutral rake and with an average chip thickness of 0.2 mm.

Many of the smaller milling machines still in use today have maximum spindle power of 4 kW and this limits them to cutters of no greater than 50 mm diameter. This lack of power in these smaller machines does not allow hardmetal cutters to show up to advantage against high speed steel tools and this has been a hindrance in the development of the market for hardmetal indexable insert slot drills and end mills. However, with the introduction of machining centres power availability is not a problem and hardmetal indexable insert cutters are gaining ground. TiN coated high speed steel cutters also need more power than that for uncoated high speed steel if they are to perform effectively. TiN coated high speed steel cutters can run at twice the speed of uncoated high speed steel tools and hardmetal can work satisfactorily at four times the speed of uncoated high speed steel. For a given design of cutter the power requirement is directly related to the feed rate which in turn is directly related to the cutting speed.

Rigidity is always an important factor in milling but becomes even more important as more power is consumed. The rigidity of the system is dependent on the machine, its condition, on the mounting of the cutting tool, on the configuration of the workpiece and finally on the holding devices used to retain the workpiece in position.

In order to combat the possibility of vibration occurring indexable insert milling cutters are designed to have unequal tooth spacing where possible. When milling aluminium and its alloys the optimum cutting speeds are very high and it is very important that the cutters should be balanced in order to achieve the high quality surface finish which is usually desired.

If the rigidity of the fixturing will not allow climb milling to be carried out and conventional milling has to be adopted then sharp cutting edges are desirable in which case PVD coated hardmetal indexable inserts are preferred rather than CVD coated inserts. Cermets with sharper edges should also be considered and especially with stainless steels can perform well in finishing operations. Sharper edges are also better for very fine finish machining. They minimise any tendency to produce burrs and are preferable for milling thin walled components.

Studs and clamps are still a commonly used method for holding workpieces. On CNC machines these can interfere with the free movement of the tooling around the workpiece. Hydraulic holding systems and strong magnets are suitable alternatives. Vacuum clamping is also used in the case of aluminium.

5.4.5 CUTTING PARAMETERS

Assuming that the method of milling and the style of milling cutter has already been decided upon two basic cutting parameters need to be considered. These are firstly the cutting speed, i.e. the linear speed of the cutting edge as it passes over the workpiece, and secondly the feed rate.

The cutting speed is quoted in metres per minute and is usually expressed as V_c where:

$$V_c = \frac{\text{cutting diameter (mm)} \times \pi \times \text{cutter revs per min.}}{1000}$$

The feed rate determines the amount of metal each tooth will remove for each revolution of the cutter. The feed per tooth is a critical factor and together with the cutting speed it is used to nominate the conditions for milling a particular workpiece with a specified cutting material in a particular milling operation. The feed rate is expressed as V_f and is quoted in mm per minute according to the formula:

$$V_f = \text{feed per tooth (mm)} \times \text{no. of teeth} \times \text{cutter revs per min.}$$

There is a recommended cutting speed range for each type of workpiece material dependent on the cutting materials being used. The less difficult the material is to machine then the higher the cutting speed which can be employed. Similarly there are recommended feed per tooth values and these are used to calculate V_f by using the formula above.

Cutting speed ranges, V_c, with corresponding values of feed per tooth, f_z, are quoted below. The values relate to the style of cutter and for the most widely used workpiece materials and the most commonly employed cutting materials.

End Mills

	HSS	TiN coated HSS	Hardmetal	Coated hardmetal
Low C steels				
V_c	25–35	50–60	80–150	120–220
f_z	0.09–0.07	0.09–0.07	0.2 –0.1	0.2 –0.1
Alloy steels				
V_c	20–25	30–35	60–120	100–150
f_z	0.09–0.07	0.09–0.07	0.2 –0.1	0.2 –0.1
Tool steels (annealed)				
V_c	10–12	15–20	60–100	80–120
f_z	0.05–0.04	0.05–0.04	0.18–0.1	0.18–0.1
Stainless steels				
V_c	10–12	15–20	70–130	80–160
f_z	0.05–0.04	0.05–0.04	0.18–0.1	0.18–0.1
Titanium alloys				
V_c	10–15	–	40–60	–
fz	0.04–0.02	–	0.1 –0.07	–
Grey cast iron				
V_c	20–30	50–60	60–90	80–130
f_z	0.09–0.07	0.09–0.07	0.3 –0.2	0.3 –0.2
Ductile cast iron				
V_c	20–30	50–60	50–80	90–120
f_z	0.09–0.07	0.09–0.07	0.12–0.06	0.12–0.06
Aluminium and Al alloys				
V_c	60–70	–	250–500	–
f_z	0.1 –0.08	–	0.3 –0.2	–

Slot Drills

The values for slot drills are as those for end mills but slot drills are not suitable for titanium alloys, also coated hardmetal is not normally proposed to be used on aluminium alloys.

With end mills and slot drills the overhang of the cutter should be reduced to a minimum. The cutter should be the shortest possible.

Porcupine Cutters

Porcupine cutters have either brazed hardmetal teeth or clamped indexable insert hardmetal teeth. A characteristic of these cutters is that the hardmetal teeth are helically staggered and offset in relation to the next row of teeth in such a manner that two adjacent rows provide a complete cutting edge. This means that these tools can operate at very high metal removal rates for roughing and can also work on relatively low powered machines.

The effect of having a helical cutting edge is that on deeper cuts a cutting edge is always in contact with the workpiece and this makes for more uniform cutting and reduced vibration.

The ranges of cutting speed and feed per tooth are given in the table overleaf for both uncoated and coated hardmetal.

Face Mills

The points which are made below about face milling apply particularly to indexable insert face milling cutters.

The way the rake face enters the cut is governed by the axial rake and the radial rake built into the insert seating. Figure 72 explains what is meant by axial and radial rakes. In this illustration the insert shown is raked negatively with respect to the axis of the cutter and also negatively with respect to the radius of the cutter.

Double negative rake geometry as shown in Figure 72 is used for cast iron milling. It results in a very strong cutting geometry because the initial point of contact between the workpiece and the indexable insert is back from the cutting edge. The disadvantage of double negative geometry is that it generates very high cutting forces and rigid conditions must exist.

Double positive geometry produces the lowest cutting forces and therefore uses much less power. It is suitable for unstable conditions but

	Hardmetal	Coated hardmetal
Low C steels		
V_c	90–170	120–250
f_z	0.3 –0.15	0.3 –0.15
Alloy steels		
V_c	70–130	100–150
f_z	0.3 –0.1	0.3 –0.1
Tool steels (annealed)		
V_c	70–110	90–150
f_z	0.3 –0.1	0.3 –0.1
Stainless steels		
V_c	70–130	80–160
f_z	0.3 –0.12	0.3 –0.12
Titanium alloys		
V_c	40–60	–
f_z	0.15–0.07	–
Grey cast iron		
V_c	70–110	80–120
f_z	0.3 –0.12	0.3 –0.12
Ductile cast iron		
V_c	60–80	80–120
f_z	0.3 –0.12	0.3 –0.12
Aluminium and Al alloys		
V_c	250–500	–
f_z	0.6 –0.3	–

the cutting edges are weaker and the initial point of contact between the workpiece and the insert is right at the cutting corner. High double positive geometry is ideal for milling aluminium.

Axial positive and radial negative rake geometry produces low cutting forces with a relatively strong cutting edge. It is an excellent all round geometry for face milling cutters.

Another factor which influences face milling operations is the angle at which the indexable inserts in the cutter enter the workpiece. Figure 73 shows the popularly used entry angles and also the development of the cutting forces which are produced by adopting these entry angles.

A 45° entry angle gives balanced axial and radial cutting forces. A 75°

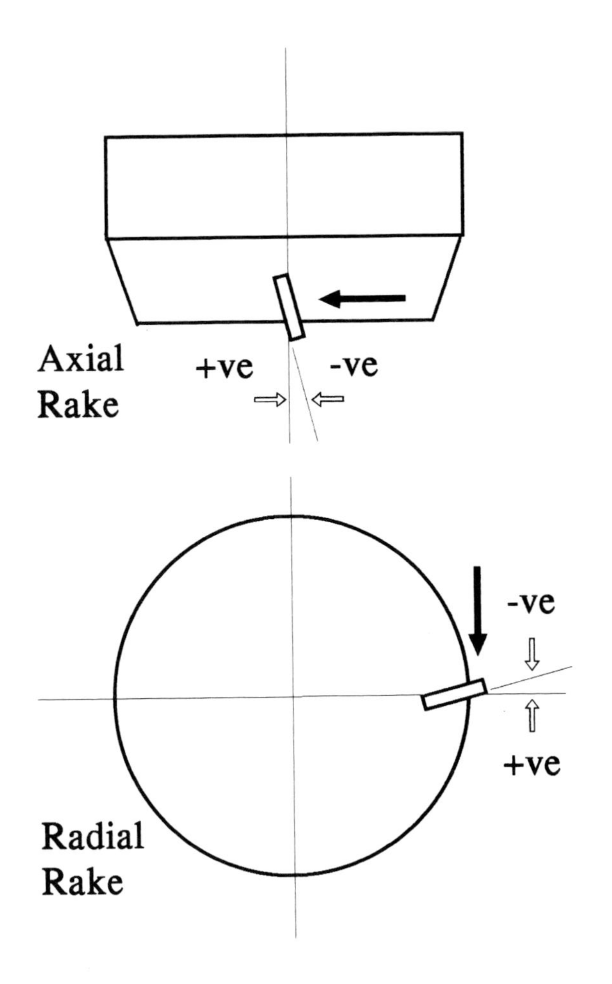

Fig. 72 Axial and Radial Rake Angles in Face Milling Cutters

entry angle allows a deeper cut to be taken but the radial cutting force is increased and this can be a disadvantage where weaker conditions exist.

An entry angle of 90° is only used when a 90° shoulder is required. Its disadvantage is that it results in the highest radial cutting forces.

The bottom diagram in Figure 73 illustrates the situation when a round insert is used. In this case the cutting edges are very strong and give an advantage with materials which are difficult to machine. The power requirement is high and stable conditions are needed. If smaller depths of cut are used then very high axial cutting forces are generated.

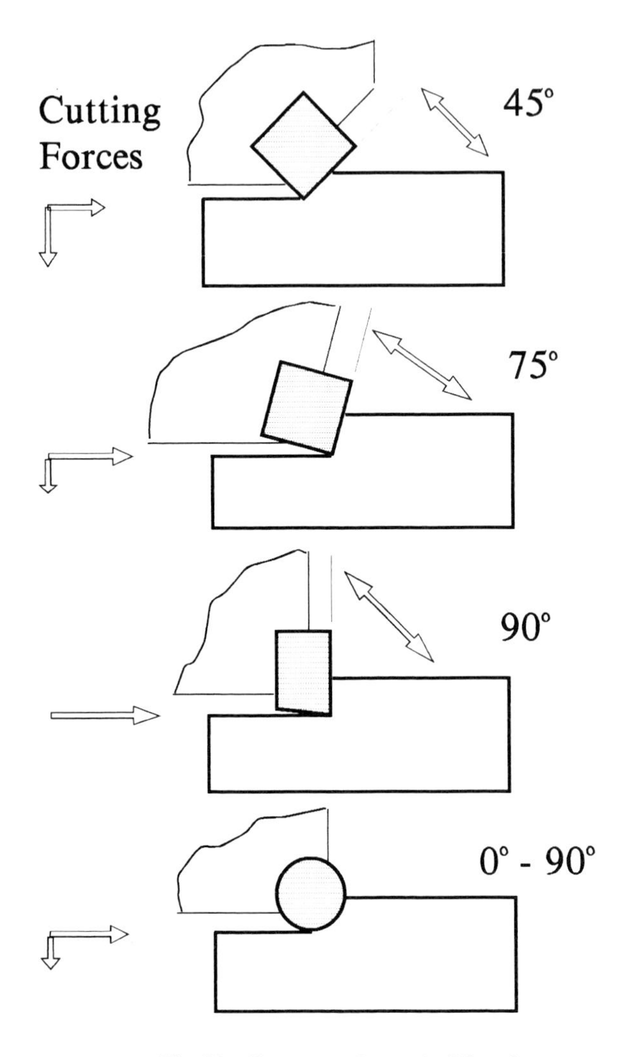

Fig. 73 Entry Angles on Milling Cutters

The entry angle also affects the chip thickness for a given feed per tooth. In Figure 74 the same feed per tooth, fz, is being applied to a 45° entry angle cutter and also to a 90° entry angle cutter. The smaller the entry angle then the thinner the chip which is produced for the same feed per tooth value.

When face milling, the position of the centre line of the cutter in

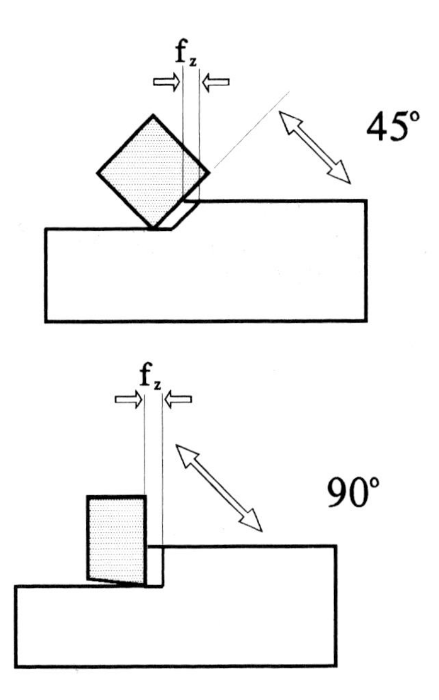

Fig. 74 Entry Angles and Chip Thickness when Milling

relation to the centre line of the workpiece is an important factor. If these two centre lines coincide and in doing so cause an equal amount of material to be removed on each side of the centre line of the milling cutter then vibration is likely to occur because the direction of the resultant cutting force will be varying either side of the centre line.

The corner configuration of a face milling indexable insert will affect the surface finish produced on the workpiece. Figure 75 illustrates that an insert with a radius at the corner will produce a rough surface. If the cutting speed is increased and a lower feed is used then a slightly better surface finish will result but reducing the tooth load to a low value will give poor cutting edge life.

An insert with a cutting facet at the corner with an adjacent parallel land will produce a much better surface finish. However, this finish will be influenced by the axial run out of the indexable inserts when fitted into the milling cutter.

The best way to produce a good surface finish is to employ a wiper insert. One of the teeth in the face milling cutter is replaced by a wiper

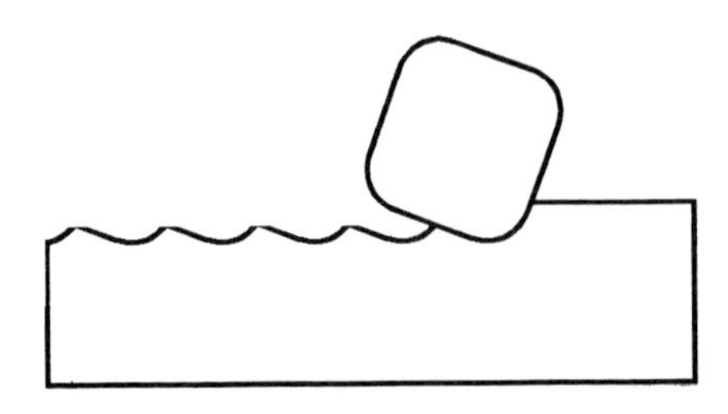

Insert with radius at corner

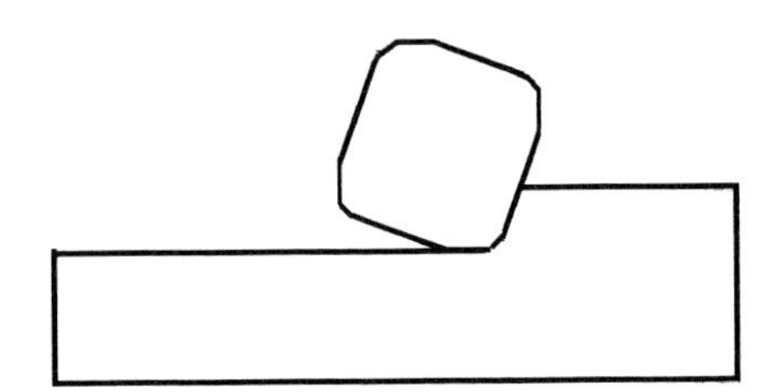

Insert with parallel facet

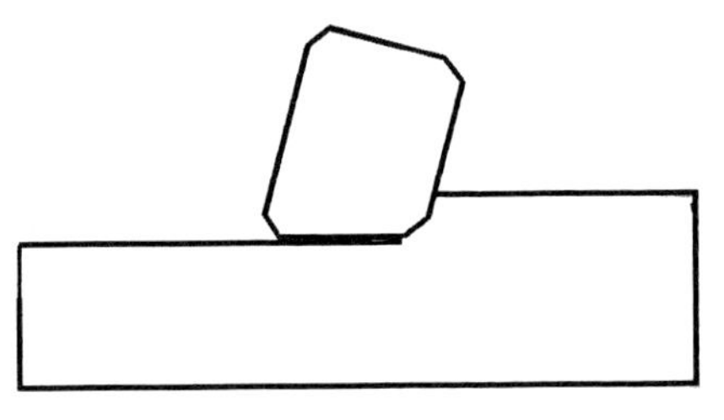

Wiper insert

Fig. 75 Surface finish v. Indexable Insert Geometry in Milling

insert which has a longer wiper edge. This will give a good surface finish even though conditions may tend to be unstable.

Cutting speed and feed per tooth values for using face mills fitted with indexable inserts are given in the table opposite.

Most face milling operations are best performed without coolant. The exceptions are heat resisting alloys, titanium alloys and aluminium alloys. With finishing cuts on stainless steel coolant can prevent smearing.

	Hardmetal	Coated hardmetal	Cermet
Low C steels			
V_c	100–200	150–250	250–450
f_z	0.4 –0.1	0.3 –0.1	0.2 –0.05
Alloy steels			
V_c	90–150	120–200	200–400
f_z	0.4 –0.1	0.3 –0.1	0.2 –0.05
Tool steels (annealed)			
V_c	60–120	100–160	–
f_z	0.4 –0.1	0.3 –0.1	–
Stainless steels			
V_c	80–150	100–200	200–400
f_z	0.4 –0.1	0.3 –0.1	0.2 –0.05
Titanium alloys			
V_c	20–80	–	20–80
f_z	0.2 –0.1	–	0.1 –0.05
Grey cast iron			
V_c	90–170	150–250	80–150
f_z	0.3 –0.1	0.3 –0.2	0.4 –0.1
Ductile cast iron			
V_c	100–150	150–250	80–150
f_z	0.4 –0.1	0.3 –0.2	0.4 –0.1
Aluminium and Al alloys			
V_c	500–2000	–	–
f_z	0.4 –0.1	–	–

Polycrystalline diamond tipped indexable inserts are used for high productivity and to reproduce an excellent surface finish on aluminium and its alloys. Cutting speeds of the order of up to three times those used for hardmetal are employed for the same feed per tooth. This requires a cutter with a coarse tooth pitch. When face milling aluminium alloys with silicon contents of 12.5 – 13.5% PCD tipped cutters can run at 2000 m min^{-1} and at these high cutting speeds attention must be paid to clearing away the chips from the cutting edges.

When face milling heat resisting alloys a round insert provides the maximum edge strength. It also limits the chip thickness and this is important in the case of heat resisting alloys. The feed should be reduced to produce a good surface finish. The axial cutting force is increased when milling with round inserts and so to counteract this the machine must be rigid. The average chip thickness should be between 0.1 and 0.12 mm.

In the case of titanium alloys positive geometry is needed. Positive axial rake reduces the cutting force and reduces the work hardening effect which occurs when machining titanium alloys. The feed should be reduced to achieve a good surface finish.

A close pitch cutter should be chosen for both heat resisting alloys and titanium alloys. The chip thickness is a limiting factor and the number of teeth determines the metal removal capacity.

Flood coolant is advised and in the case of titanium alloys a mist coolant is recommended.

5.4.6 MILLING WITH CERAMICS & CBN

Although ceramics and CBN are not standard choices for milling there are particular cases where they perform excellently.

Ceramics

Ceramics are being used in the automotive industry to machine components such as gear box cases and cylinder heads made from grey cast iron. Rough face milling is being done using negative rake silicon nitride indexable inserts. The excellent thermal shock resistance of silicon nitride makes it suitable for rough milling of cast iron and with the negative rake geometry it is able to take interrupted cuts and cope with unfavourable casting skins. Cutting speeds of 500 to 600 m min^{-1} and feed per tooth values of 0.2 to 0.4 mm are quoted.

CBN

The milling of the slide ways of machine tool beds is one of the successful applications for cubic boron nitride. It has been possible to make considerable reductions to grinding time and in some cases grinding has been eliminated because the accuracy and quality of the surface finish achieved by CBN is outstanding.

These machine tool beds are made from cast iron (meehanite) which is then induction hardened to a minimum of 50 HRC. Round CBN indexable inserts, negatively inclined in the face milling cutter are normally used with cutting speeds from 350 to 400 m min^{-1} and feeds of the order of 0.4 mm per tooth.

5.4.7 HARDMETAL INSERT PROBLEMS

The following chart may be of help in overcoming some of the problems which can occur during milling with hardmetal indexable inserts.

Identifying problems and possible courses of action when milling with hardmetal indexable insert cutters

	Thermal cracking	Insert chipping	Rapid flank wear	Built up edge	Crater wear	Plastic deformation
Reduce the cutting speed	*		*		*	*
Increase the cutting speed		*		*		
Reduce the feed per tooth		*			*	*
Increase the feed per tooth			*	*		
Use a more wear resistant grade			*			*
Use a more crater resistant grade					*	
Use a tougher grade	*	*				
Do not use coolant	*					

Standards

At the end of the book is a list of standards which apply to Hard Material cutting tools. Apart from the indexable insert standard which has already been described in earlier chapters (ISO 1832) the following standards relate to milling.

Dimensions for milling	ISO/3365
Dimensions for milling, wiper	ISO/3365 Pt3
Dimensions of end mills, parallel shanks	ISO/6262 Pt1
Dimensions of end mills, morse taper shanks	ISO/6262 Pt2
Dimensions of face mills	ISO/6462
Dimensions of side and face mills	ISO/6986
Designation of bore type cutters	ISO/7406
Designation of shank type cutters	ISO/7848
Dimensions of plain parallel shanks	ISO/3338 Pt1
Dimensions of flatted parallel shanks	ISO/3338 Pt2
Brazed helical end mills, parallel shanks	ISO/DP10145 Pt1
Brazed helical end mills, 7/24 taper shanks	ISO/DP10145 Pt2
Dimensions of solid hardmetal end mills	ISO/CD10911
Designation of solid hardmetal end mills	ISO/CD11529
Tool life testing, face milling	ISO/8688 Pt1
Tool life testing, end milling	ISO/DIS9766

5.5 DRILLING

The dominant tool for making holes in metal components is the high speed steel twist drill. About 80% of all metal drilling is done using high speed steel as the cutting material. Such is the popularity of DIY (Do It Yourself) that most homes will have an electrically driven power drill and a few HSS twist drills. It is therefore assumed that it is unnecessary to include an illustration of a twist drill in this book. As well as being made from high speed steel, twist drills are also produced with brazed hardmetal tips and from solid hardmetal itself. TiN coating is also applied both to high speed steel drills and to solid hardmetal drills and so a variety of cutting materials is available for consideration.

Holes with large length/diameter ratios can be made using Gun Drills. As their name implies, one of the purposes of this type of drill is to produce the holes in rifle barrels. These drills have a brazed hardmetal tip and will be fully discussed later in this chapter.

The third type of drill which will be described is the so called Short Hole Drill. This is a popular tool for use in CNC machines and is fitted with specially shaped hardmetal indexable inserts.

Twist drills, gun drills and short hole drills are the three most popular types of drilling tool in use. The following comment relate to their design and application.

5.5.1 TWIST DRILLS

A standard high speed twist drill is made from a round bar which has two helical flutes ground in part way along its length. The main purpose of these flutes is to conduct the chips produced in the operation out of the hole and into the clear space beyond the hole. They also serve to allow coolant to pass down to the cutting area. The shank of the drill is then either a plain cylinder or will have a Morse taper. They are then known as straight shank drills or taper shank drills respectively.

In the case of the larger drills the shank is often made from a less expensive steel which is butt welded to the high speed steel cutting end of the drill. The point of the drill is ground to suit the workpiece material being drilled.

The cutting speed of the drill is usually defined as the rate at which the periphery of the drill moves in relation to the workpiece being drilled and is measured in metres per minute.

The following table gives a suggested range of cutting speeds for high speed steel twist drills when machining a variety of workpiece materials:

Aluminium and Al Alloys	30–45 m min^{-1}
Brass (cast)	60–90
Brass (wrought)	40–60
Copper	30–60
Grey Cast Iron	15–30
Ductile Cast Iron	10–20
Hard Cast Iron	8–10
Free Machining & Low C Steels	25–40
Alloy & Medium to High C Steels	10–15
Tool Steels (soft condition)	6–10
Stainless Steels	6–10
Titanium and Ti Alloys	3– 8
Plastics and Non Metallics	15–35

The feed of the drill depends not only on the workpiece material but also on the diameter of the drill.

A general guide of feed ranges for steps of drill diameter is given below:

Drill diameter (mm)	Feed range (mm)
up to 3	0.02 to 0.05
3 to 6	0.05 to 0.1
6 to 12	0.1 to 0.2
12 to 25	0.2 to 0.4
over 25	0.4 to 0.6

The use of cutting fluids when drilling is strongly recommended. During drilling the chips can heat up to a point where they will weld on and stick to the tool and this is particularly likely to happen when drilling steel. If this occurs the tool will fail very quickly. A good supply of cutting fluid will keep the temperature down and deter any tendency of chips sticking to the drill.

The included angle of the cutting point is varied for different workpiece materials. For general purpose drills the conventional point angle is 118°. When drilling more difficult to machine workpiece materials such as stainless steel and hard materials the point angle is increased to 135°. Drills for plastics and other non metallics use a smaller point angle of 90° and less.

Coating high speed drills with Titanium Nitride significantly increases their productivity by permitting them to operate at higher cutting speeds and increased feeds. Cutting speed increases of 50% and more are typical of the improvement in performance from the gold coloured TiN coating. These coated drills are also able to operate on a wider variety of difficult to machine materials.

Regrinding of high speed steel drills should be done in accordance with the instructions laid down by the manufacturers. This is probably one of the worst areas of tool reservicing and it should be noted that the reputable drill suppliers provide good information on drill application and on drill regrinding.

CNC machines have helped to increase the use of solid hardmetal drills. Two further developments have added to this increase and they are the introduction of ultra fine grain WC hardmetal cutting materials and the development of solid hardmetal drills with spiral coolant holes. By combining a TiN coating with internal coolant supply via the spiral coolant holes hardmetal drills can give an outstanding performance on the very difficult to machine materials.

The use of ultra fine grain WC, of the order of 0.5 micrometres, not only brings about higher wear resistance but also results in considerable improvement in bend strength over previously used fine grain WC hardmetals with grain sizes in the region of 0.75 to 1.0 micrometres. By employing these 'stiffer' hardmetals drill deflection on starting to enter the workpiece is considerably reduced and this together with the more compatible CNC machines means that drill breakage with hardmetals is no longer a major concern.

The grades of hardmetal offered lie in the ISO K10 – K20 application groups and for steel drilling the ISO P30 group of grades is available but

TiN coating on the plain Co-WC grades avoids the necessity of using the crater resistant P group of hardmetal grades.

British Standard BS 328 entitled 'Twist Drills and Combined Drills and Countersinks' covers dimensions and tolerances and is referred to in manufacturers product literature.

5.5.2 GUN DRILLS

Gun drilling is a very precise hole making process. Straight holes with a good finish and to close dimensional tolerance can be produced in a broad diameter range from less than 3 mm to around 50 mm. With the smaller diameters lengths up to 100 times the diameter are achievable and in the larger diameters lengths of more than one metre can be drilled in one pass.

A gun drill consists of a solid piece of hardmetal in the case of the smallest sizes, or more usually, a brazed hardmetal tip connected to a shank which locates into a driver. Coolant is passed through the driver and through the shank into the hardmetal tip which has an exit hole or holes located adjacent to the cutting edge. The smaller diameter drills have only one hole and the larger ones have two. The performance of the drill very much depends on the volume and pressure of the coolant which emerges at the cutting end of the drill. A 'V' shaped flute in the shank allows the chips which are formed to be flushed out by the pressurised coolant. Figure 76 is a photograph of a gun drill.

To commence cutting, the head of the drill must be fed through an accurately ground bush which must fit tightly to the face of the workpiece being drilled and must not move as drilling commences. Alternatively a predrilled hole can be used as a starting situation.

Figure 77 illustrates the head of a gun drill. The cutting edge is in two parts as can be seen on the right side of centre in Figure 77. This profile produces two thin converging chips which have a width equal to half the drill diameter and which break into manageable lengths on contact with each other. The periphery of the hardmetal tip becomes a burnishing pad which generates a good surface finish in the hole and also promotes good accuracy. Once the hole is started it takes the place of the bush and the drill becomes self piloting.

The best results are achieved with machines which are specially designed for gun drilling. Such machines must have good stability so that the possibility of vibration is minimised and they must have sufficient power to perform the drilling operation. Spindle speeds must be suitable

Fig. 76 A Hardmetal Tipped Gun Drill

for the diameter of drill to be used and a mechanism which can deliver a constant feed, preferably stepless, is essential. Finally and most important an effective coolant supply must be incorporated. Conventional machines can be modified provided the points mentioned above are observed.

The workpiece must be securely held in correct alignment in the machine. When drilling long workpieces, such as rifle barrels, steadies should be used on both the workpiece and on the drill shank.

As with twist drills, the feed is related to the diameter being drilled. The table opposite gives some guidelines of cutting speeds and feeds for a variety of workpiece materials:

Fig. 77 Cutting Head of a Hardmetal Gun Drill

Workpiece material	Cutting speed m min⁻¹	Drill diameter 1–3	3–6.3	6.3–12.5	12.5–35
		Feed mm/rev			
Aluminium and Al alloys	65–300	0.005–0.015	0.005–0.04	0.02 –0.07	0.03 –0.15
Brass, bronze, copper	65–300	0.005–0.015	0.005–0.04	0.02 –0.07	0.03 –0.15
Gray cast iron	60– 90	0.004–0.01	0.005–0.03	0.01 –0.07	0.03 –0.19
Ductile cast iron	70– 90	0.005–0.01	0.008–0.03	0.02 –0.07	0.05 –0.19
Unalloyed, low C steel	60–120	0.003–0.01	0.005–0.03	0.015–0.055	0.02 –0.11
Low alloyed, medium C steel	40–120	0.003–0.01	0.004–0.03	0.01 –0.055	0.02 –0.11
Stainless steel	40– 90	0.003–0.008	0.004–0.025	0.01 –0.04	0.02 –0.1

Hardmetal tipped gun drills are being coated with TiN by the PVD process which is carried out at temperatures below the melting point of the braze material. These PVD TiN coated gun drills can operate at higher speeds than standard gun drills and still give an excellent performance after regrinding.

The life of the cutting edge before regrinding becomes necessary averages between 10 and 30 metres of cutting length depending on the workpiece material being machined. The number of regrinds which can be expected is between 15 and 20. Grinding must be accurately carried out reproducing the original cutting geometry. The quality of the ground edge plays a large part in the performance of the drill and should be carefully controlled. Any roughness in the ground finish will cause immediate breakdown of the cutting edge to commence and tool life will be very short.

The following points summarise the role of gun drills in machining:

- They are economical tools for precision drilling of small holes.
- Fine tolerances and a good finish are obtained.
- They can be used for both short and long holes.
- Cutting speeds can be 4 to 5 times those for high speed steel twist drills.
- When chip breaking is difficult they can be useful tools.
- They can be used for drilling very hard materials.

5.5.3 SHORT HOLE DRILLS

Short Hole Drills as their name implies are used to produce comparatively short holes up to a recommended safe length of 2.5 times the diameter of the hole being drilled. They employ clamped hardmetal indexable inserts as their cutting material and holes which can be produced are typically in the range 20 to 55 mm diameter.

The indexable inserts which are most commonly used are the 6 sided trigon inserts designated by the letter W in the ISO designation system. They are secured by means of a screw and so there are no overhead clamping elements to obstruct the flow of chips within the hole.

As with gun drilling, coolant plays a large part in the successful operation of the drill. Holes pass through the shaft of the drill and emerge at the end of the drill close to the cutting edges. The flutes of the drill run parallel to its axis which gives the tool maximum resistance to torsion and bending. The straight flutes also provide the shortest path for evacuation of the chips produced. The coolant is supplied under pressure and as well as cooling and lubricating the cutting area it flushes away the chips so that they do not damage the cutting edges of the indexable inserts or the surface of the hole being drilled.

The prime function of this type of drill is its high penetration rate . It employs two cutting inserts, one central and one peripheral which over-

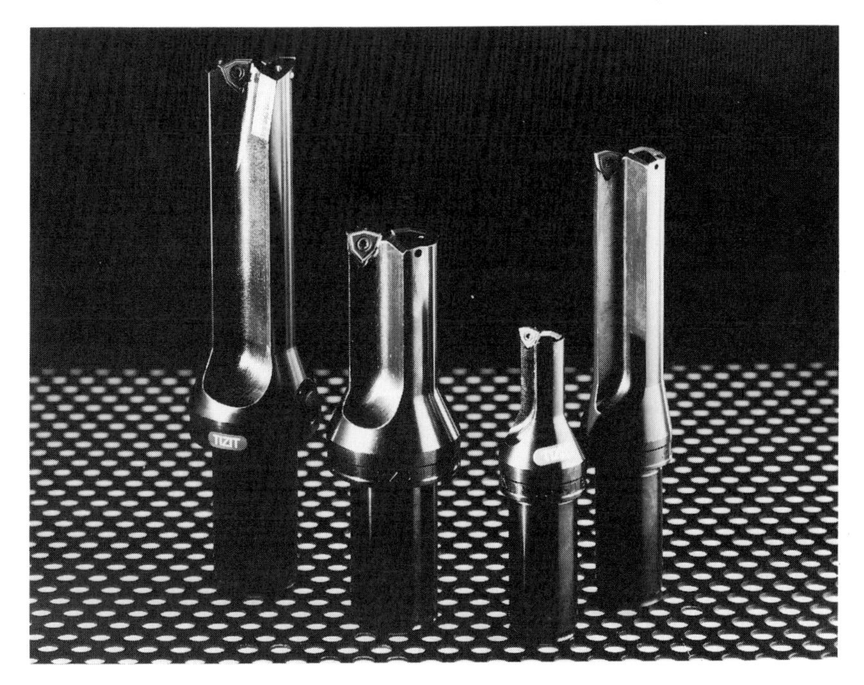

Fig. 78 Short Hole Drills with Hardmetal Indexable inserts

lap to create the required cutting action. The shank of the tool can be supplied so that the tool is suitable for most types of machine and shank type holding system. It can be used as a rotating drill or as a stationary tool with the workpiece rotating. Figure 78 shows short hole drills with different diameters. The screwed on trigon hardmetal indexable inserts can be clearly seen also the large straight flutes which have good chip clearance capability.

The advantages of short hole drills can be summarised as:

- High cutting speeds
- Large feeds
- Faster machining times
- Longer tool life
- No need to pre-centre
- Lower feed forces than with high speed steel drilling
- Cutting edges can be indexed
- No regrinding of cutting edges
- Hardmetal grades can be chosen to suit workpiece material
- Controlled chip breaking by control grooves in inserts

- High level of precision and accuracy of hole
- Short machine down time due to quick changing of tools

The radial forces which occur during drilling normally push the drill sideways but by specially arranging the position of the inserts in the short hole drill the radial forces are balanced out. This prevents run out and improves the surface quality over the whole depth of the bore.

The 'W' style trigon inserts have six sides in three pairs. Each pair of sides forms one of the cutting edges of the indexable insert and has an included angle of 156°. A clearance angle of 7° is provided in the insert when set at neutral rake in the tool. An insert can be indexed three times before it must be discarded.

When using short hole drills clearance face wear must be monitored regularly. The cutting forces rise with increased clearance face wear and if an excessive amount of clearance face wear is allowed to be set up then the demands on the indexable insert may become too great and the insert could break and the tool become damaged.

If the surface to be drilled is not flat there is the likelihood that one insert is in contact with the workpiece whilst the other is still out of the cut. This will give rise to an out of balance condition which will lead to run out of the tool if no action is taken. In these circumstances it is advisable that the feed is reduced to at least half that normally recommended.

One important point concerning safety is that as the final part of the hole is being drilled with this type of tool a disc of workpiece material comes away as the inserts break through. A cover should be placed over the operation so that this disc cannot fly off and cause injury or damage.

If a predrilled hole exists it must not have a diameter larger than one quarter the finished size otherwise the short hole drill will deflect because the two cutting inserts will not be equally loaded and an out of balance situation will occur.

A table of recommended ranges of cutting speeds and feeds for different workpiece materials and drill diameters is given opposite:

Drilling conditions which are likely to exist can be represented by the following three statements:

Bad conditions – Use lower speeds
Normal conditions – Use moderate speeds
Good conditions – Use high speeds

The grades of hardmetal which are suitable for these conditions will

Workpiece material	Cutting speed m min^{-1}	Drill diameter Up to 25	25–30	30–40	Over 40
		Feed mm/rev			
Aluminium and Al alloys	150–375	0.08–0.11	0.11–0.17	0.17–0.27	0.17–0.27
Brass, bronze, copper	80–160	0.09–0.15	0.09–0.15	0.15–0.25	0.15–0.25
Gray cast iron	80–180	0.09–0.15	0.15–0.22	0.15–0.22	0.18–0.3
Ductile cast iron	90–200	0.11–0.18	0.14–0.22	0.17–0.27	0.18–0.3
Unalloyed, low C steel	100–250	0.04–0.12	0.09–0.19	0.11–0.2	0.14–0.25
Low alloyed, medium C steel	90–250	0.08–0.12	0.09–0.16	0.11–0.2	0.14–0.22
Tool steels – soft	100–220	0.08–0.12	0.09–0.18	0.11–0.22	0.14–0.25
Tool steels – hardened	90–200	0.08–0.12	0.09–0.15	0.11–0.17	0.12–0.2
Austenitic stainless steel	70–150	0.04–0.12	0.1 –0.16	0.11–0.18	0.11–0.18
Ferretic stainless steel	90–190	0.04–0.12	0.1 –0.16	0.11–0.18	0.14–0.18

fall into the ISO application groups P40, P30 and P20 respectively for steels which form a crater and K20, K20 and K10 respectively for all other workpiece materials.

Coated hardmetals perform very well in short hole drilling tools and when using them the question of crater or no crater does not apply.

In some cases it can be advantageous to use a tougher grade of hardmetal for the centre indexable insert and a more wear resistant one for the peripheral insert.

5.6 MODULAR TOOLING & QUICK CHANGE SYSTEMS

With the ever increasing pressure to reduce manufacturing costs and to increase productivity machine tools have been the focus of attention. If the output per hour from the machine can be increased significantly then the consideration is – at what cost? In earlier days supervisors concentrated on the cutting tool and tried to reduce costs and increase output by trying to apply cutting tool materials which would give the longest cutting edge life under the same machining conditions. At that time almost all of the tools were removed from the machine for regrinding when they had become worn. The time required to remove and replace the tool was itself a loss to production but the resetting time which had to be taken to ensure that the dimensions of the component were correct was an equally negative factor. Even if the machine was capable of running at increased speeds and feeds the wear on the tools would accelerate, the frequency of changing would increase and the cost of the lost time involved would be unacceptable.

With the introduction of indexable insert tooling changing a cutting edge became a short operation and coupled with the advances in cutting materials has had a profound effect on productivity. Coatings, cermets and ceramics have enabled cutting parameters to be raised significantly and the developments in machine design have kept pace with this. However, the machines of today which have evolved are very expensive and the amortisation of the machine is now by far the greatest part of the total cost per hour for operating the machine.

Except for CBN and PCD the cost of the cutting edge is a very minor element in the total cost of the component being produced. Because the cost of the cutting edge is not a major factor, cutting parameters should be optimised so that the maximum sensible speed, feed and depth of cut are employed. If this is done the machine will give the maximum productivity whilst it is cutting and the only loss to production will be the time the cutting edge is not in contact with the workpiece. Thus, assuming that cutting materials and cutting parameters have been optimised, in order to reduce 'floor to floor' times the savings must come from either the time involved in installing and removing the workpiece or the time taken for inward and outward travel of the cutting tool before and after performing the cutting operation or the time taken to index and change the cutting edge which could also involve presenting an alternative tool geometry to the workpiece.

The first two of these possible ways of saving are the concern of the machine manufacturers but the one which involves the cutting edge is the concern of the cutting tool suppliers. Thus the designers have concentrated their efforts in developing modular tooling and tool change systems.

A modular tooling assembly can be said to be made up from various components such as a basic holder, extension pieces to lengthen the tool, reducers to take different diameter heads and of course the cutting head itself. All the components should be interchangeable and should all have a common coupling system. In this way, for any specific application, the relevant components can be assembled into the required tool.

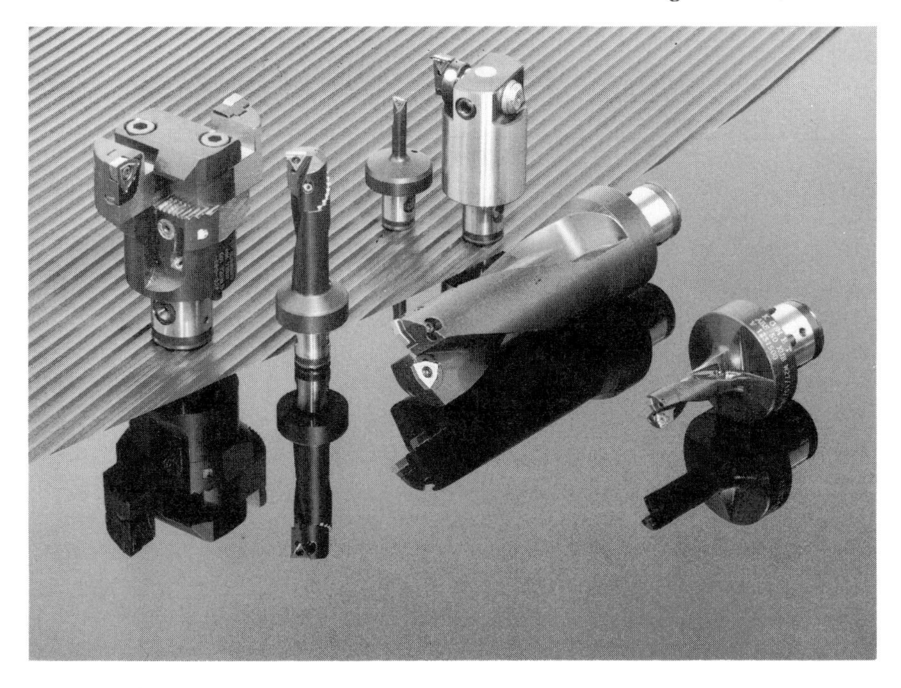

Fig. 79 Drilling/Boring Tools with Interchangeable Connectors

Modular tooling systems also operate as tool change systems and Figure 79 shows a selection of drilling and boring tools with interchangeable connectors and which form part of a well established modular system. These tools fit into common basic holders which would be mounted in the machine.

The ideal system should be capable of being applied to both stationary and rotating tools and Figure 80 illustrates milling cutters which have the same connecting system as that used in Figure 79.

In the case of milling or drilling machines a basic holder can be mounted into the spindle which can then accept adaptors, mountings and chucks which can clamp the normal straight shank and taper shank slotting cutters, end mills and drills etc. and a selection is shown in Figure 81. These devices have the same interchangeable connector as the milling cutters and drilling and boring tools shown in Figures 79 and 80.

Tool changing systems can be applied to manually operated machines, NC lathes, CNC machining centres and flexible machining cells. An extremely important point about these systems is that the cutting

Fig. 80 Milling Cutters with Interchangeable Connectors

edge indexing and setting can be done away from the machine. The tool change is quickly done with high repeatability. The connection must be positive and backlash free ensuring high rigidity between the tool head and the holding unit. Most suppliers offer the possibility of a chip being installed in the cutting head which gives the possibility of read only or read/write data being carried by the cutting head. This chip can then receive, store and deliver information as the tool is being serviced away from the machine.

With any tooling which is made up by assembling a number of components it is vital that each connection ensures that accuracy and rigidity are maintained throughout. This presents a formidable design task and it is not surprising that cutting tool manufacturers have come up with solutions which are different from those of their competitors. The Block Tool System, FTS, ABS, Varilock and UTS are tool change and modular system designs which immediately come to mind. Three of the major cutting tool producers have adopted the UTS (Universal Tooling System) as their choice and as this system is the one best known to the author it will be used to illustrate the major features of a modular system. However, this in no way implies that other systems cannot perform equally well.

Fig. 81 Chucks, Adaptors and Mountings for a Modular System

The conditions which should be satisfied in any tool changing modular system can be stated as:

- High static and dynamic rigidity.
- Repeatability of cutting edge position on tool changing.
- A variety of assembly sizes.
- Can be installed on all types of conventional machines.
- Flexibility in installation e.g. extensions, reducers, collets, tool heads.
- Can be used in all machining operations – turning, milling, drilling, threading etc.
- Facility for internal coolant supply.

The key to the performance of tool changing and modular systems is the viability of the coupling which takes place between the holder (the receiver) and the part being inserted into the holder. The strength and rigidity of the joint are vital in that no movement must take place as the cutting load comes on otherwise the accuracy of repositioning is lost and repeatability fails.

The UTS coupling system consists of two basic parts – a male and a female. The male part is the tool head or adaptor, chuck etc. and the female part is the holder into which the male is to be coupled. The male has a square shoulder which is the abutment face of the joint and from which a short taper projects out. There are two holes in the taper which are diametrically opposed to each other. The purpose of these holes is to receive two balls which are then used to draw the male forward into the female and also push out the taper to make a tight fit in a matching taper in the female. A male part is shown in Figure 82. This is a UTS parting/grooving tool head and alongside is a conventional toolholder which performs the same operation. In both these cases an additional module is incorporated which allows the blade of the tool to be changed to a different width and to carry alternative inserts but this module has no relevance to the subject under discussion in this chapter and should be ignored as the purpose of the photograph is purely to illustrate the male UTS connection.

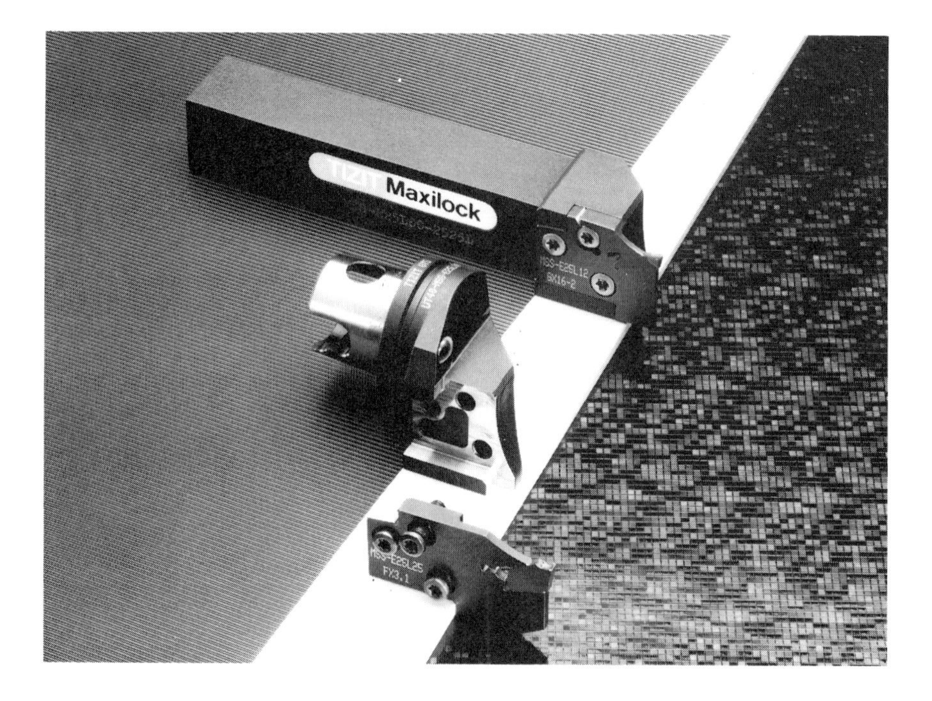

Fig. 82 A UTS System Connector for Parting and Grooving Tools

The female has a bore which is a matching taper with the male projection. It then has either a draw bar device, mounted axially, or a specially shaped pin, mounted radially, which is used to cause two balls to move outwards and enter the holes in the male taper – they are not allowed to project beyond the surface of the male taper. As the movement of the draw bar or pin is increased the force exerted by the balls elastically deforms the mating faces. The locking force used must be controlled and torque values are quoted when locking is effected by means of a screw and then the use of a torque wrench is required. In the case of hydraulic actuation a pressure gauge should be employed. The tool heads have a tensile strength of 1300–1400 N mm^{-2}.

The radial locking method is used mostly where manually operated tool changing is involved and also in the case of modular construction. It is probably the most frequently used method at the present time.

The area of use for axial locking lies mainly with automatic tool changing. In this case drum or chain or disc storage systems are used to hold the tool heads, adaptors and chucks. A tool head can be removed from these storage units for servicing whilst it is in the stand by situation and the reserve tool head can be inserted in its place.

All the tool heads, adaptors and other components which conform to the same assembly size can be coupled into the appropriately sized basic holders. Whether they have radial or axial locking methods makes no difference. The size of the coupling limits the cutting forces which can be absorbed and six sizes are offered – 25, 32, 40, 50, 63, and 80. The larger the indexable insert being used then the larger will be the size of the assembly.

Internal coolant supply is provided with the UTS system and it is claimed that an essential feature is that the coolant does not flush into the locking elements as any remnant debris in the coolant could cause accelerated wear in the locking area.

Precleaning of the parts to be coupled is essential and should be part of the procedure laid down for manual tool changing. Precleaning is built into most of the machines with automatic changing systems.

6
Practical Machining Examples

In this final part of the book an attempt is made to take the range of workpiece materials listed in the table on page 84 and present practical examples of machining with single point tools using the cutting materials most likely to be involved.

It is felt that by taking cases which are typical of what is done and by not attempting to show a cutting material performing near its absolute limit then this will be of more value to the reader.

6.1 ALUMINIUM & Al ALLOYS

Examples are given of turning with High Speed Steel, Uncoated Hardmetal and Polycrystalline Diamond.

High Speed Steel

In this instance a cast aluminium billet is being machined before being loaded into an extrusion press. The billet is 675 mm long and 250 mm in diameter. The surface quality of the vast majority of the billets as received from the continuous casting process is quite acceptable and the diameter of the billets normally falls well within the allowable tolerance band. However, the occasional billet does not reach the standard of acceptance because of surface defects but can be rectified by machining the diameter. This turning operation is not part of routine production and in this particular case is carried out by the maintenance staff as and when required. The maintenance department has a lathe which is used for repair jobs and also a selection of tools, most of which are made from high speed steel.

WORKPIECE	– A billet 250 mm diameter × 675 mm long
MATERIAL	– Aluminium 99.9% pure
OPERATION	– Turning the diameter over the full length
MACHINE	– Centre lathe
TOOL	– High Speed Steel (M2) Butt Weld No. 7
RAKE ANGLE	– + 25°
CORNER RADIUS	– 1 mm
EDGE CONDITION	– Sharp
CUTTING SPEED	– 75 m min^{-1}
DEPTH OF CUT	– 2 mm
FEED	– 0.4 mm/rev

Coolant is not used. Chip control is difficult. Chips from this very pure aluminium do not break easily. The time taken to cover the length of the billet at this speed and feed is seventeen and a half minutes. High speed steel is capable of cutting at up to nearly twice the speed chosen but operators are cautious about running large components at high speeds although in the case of aluminium weight is not so much a problem.

As already stated, chip control is difficult and a saving of a few minutes in cutting time might be swallowed up with the additional chip handling problems the faster speed would present.

Uncoated Hardmetal

Two examples are given – one is machining pure aluminium and the other is machining an aluminium alloy.

Aluminium frying pans are turned on the base using uncoated hardmetal indexable inserts which have special high rake aluminium cutting geometry.

WORKPIECE	– A frying pan – 250 mm diameter
MATERIAL	– Commercially pure aluminium
OPERATION	– Turning the base of the pan
MACHINE	– Herbert No. 8 Preoptive
TOOLHOLDER	– SCRCR 2525 M12
INSERT	– CCGT 120408 FN (aluminium geometry)
GRADE	– ISO application group K10
EDGE CONDITION	– Sharp
CUTTING SPEED	– 920 m min^{-1} (constant speed)

DEPTH OF CUT	– 3 mm
FEED	– 0.2 mm/rev
COOLANT	– Yes

Aluminium alloy wheels for automobiles are cast and then turned with uncoated hardmetal indexable inserts having the same special high rake aluminium cutting geometry. The operation described here is profile turning the rim of the wheel.

WORKPIECE	– A cast aluminium alloy wheel
MATERIAL	– An aluminium alloy containing 11% silicon
OPERATION	– Profile turning the rim of the wheel
MACHINE	– Modified Churchill Redman lathe
TOOLHOLDER	– SVJCL 3225 P16
INSERT	– VCGT 160412 FN (aluminium geometry)
GRADE	– ISO application group K10
EDGE CONDITION	– Sharp
CUTTING SPEED	– 700 m min^{-1}
DEPTH OF CUT	– Varying up to 3 mm
FEED	– 0.13 mm/rev
COOLANT	– Yes

PCD

Machining the grooves in pistons made from aluminium alloy has normally been done with a K10 ISO application group hardmetal but PCD is now being employed as well.

WORKPIECE	– A piston for an automobile engine
MATERIAL	– An aluminium alloy containing 18% silicon
OPERATION	– Machining the piston ring grooves
MACHINE	– Special purpose machine
TOOL	– Grooving blade tipped with PCD (brazed)
RAKE ANGLE	– Neutral, top of PCD tip is polished
SIDE CLEARANCE	– 6°
GRADE	– PCD
EDGE CONDITION	– Sharp
CUTTING SPEED	– 500 m min^{-1}
GROOVE WIDTH	– 2 mm
FEED	– 0.1 mm/rev

GROOVE DEPTH	– 4 mm
COOLANT	– Yes

6.2 BRASS, BRONZE, COPPER

For this group of workpiece materials, cases of machining with High Speed Steel, Uncoated Hardmetal and PCD are quoted below.

High Speed Steel

In this example a manufacturer of valves is machining cast gunmetal valve bodies. The particular operation picked out is the machining of the flanges.

WORKPIECE	– A cast gun metal valve body 150 mm diameter
MATERIAL	– BS 1400 LG2
OPERATION	– Turning and facing the flange faces of the body
MACHINE	– Herbert Preoptive – large capstan lathe
TOOL	– Butt Weld No. 29 Square Nose Turn & Face Tool
MATERIAL	– M35 HSS
RAKE ANGLE	– + 5° (maximum)
CORNER RADIUS	– 1 mm
EDGE CONDITION	– Sharp
CUTTING SPEED	– 60 m min^{-1} reducing to 35 m min^{-1} (constant r.p.m.)
DEPTH OF CUT	– 1.2 mm (maximum)
FEED	– 0.25 mm/rev
COOLANT	– Yes

Uncoated Hardmetal

A groove is to be machined in a thick walled brass tube. The tool involved is one of the 'Self Grip' design of grooving tools.
The cutting edges are required to be sharp and no advantage will result if coated hardmetal is used.

WORKPIECE	– A thick walled brass tube 90 mm diameter
MATERIAL	– MS 63 Brass
OPERATION	– Machining a groove 3 mm wide 3 mm deep
MACHINE	– Centre lathe
TOOL	– Indexable insert blade with Self Grip clamping
INSERT	– 3mm wide grooving insert
GRADE	– ISO application group K20
EDGE CONDITION	– Sharp
CUTTING SPEED	– 245 m min^{-1}
GROOVE WIDTH	– 3 mm
FEED	– 0.1 mm/rev
GROOVE DEPTH	– 3 mm
COOLANT	– Yes

The coolant assists the removal of chips from the groove.

PCD

PCD is used to machine the mouthpieces of musical instruments. Trumpet mouthpieces are made from brass. A cast brass rod is drilled then turned and bored with PCD tooling. The rough turning of the mouthpiece is detailed in the example below.

WORKPIECE	– Trumpet mouthpiece blank 12 mm O.D. × 75 mm
MATERIAL	– MS 58 Brass
OPERATION	– Rough profiling the outside diameter
MACHINE	– CNC lathe
TOOLHOLDER	– Custom built holder for a triangular insert
INSERT	– Triangular insert tipped with PCD
GRADE	– PCD
EDGE CONDITION	– Sharp
CUTTING SPEED	– 300 m min^{-1}
DEPTH OF CUT	– Varying up to 2 mm
FEED	– 0.1 mm/rev
COOLANT	– Yes

The finishing operation is made with a 0.3 mm depth of cut and a feed of 0.05 mm/rev.

6.3 GREY CAST IRON

Four cutting materials are chosen as examples for machining grey cast iron. They are Uncoated Hardmetal, Coated Hardmetal, Al_2O_3 Ceramic and Silicon Based Ceramic.

Uncoated Hardmetal

A cast iron bearing housing is to be accurately finished in the bore. The tolerance which must be achieved is +/- 0.015 mm.

WORKPIECE	– Cast iron bearing housing
MATERIAL	– BS 1452 grade 300
OPERATION	– Finish bore to the size of the bearing O.D.
MACHINE	– Combination lathe
TOOL	– Micro Bore 93° approach angle
RAKE ANGLE	– 0°
CORNER RADIUS	– 0.4 mm
GRADE	– ISO application group K10
EDGE CONDITION	– Ground – sharp
CUTTING SPEED	– 200 m min^{-1}
DEPTH OF CUT	– 0.5 mm
FEED	– 0.15 mm/rev
COOLANT	– No

A high surface finish is also required and this is achieved with the combination of feed, speed and corner radius used.

Coated Hardmetal

In this example the workpiece is a grey cast iron water pump body. The operation involved is to turn the end face of the body flange.

WORKPIECE	– Cast iron Water pump body
MATERIAL	– GG 25
OPERATION	– Facing the end flange
MACHINE	– Warner Swasey 2 AC
TOOL	– 75° approach angle
RAKE ANGLE	– 5° negative
CORNER RADIUS	– 1.2 mm
GRADE	– ISO application group K05-K15

EDGE CONDITION	–	Rounded
CUTTING SPEED	–	320 m min^{-1}
DEPTH OF CUT	–	3 mm
FEED	–	0.35 mm/rev
COOLANT	–	No

Coated hardmetal grades are usually capable of spanning a wider application range than uncoated hardmetals for this reason the grade is quoted as a band of applications rather than a single one.

The criteria for this operation is that the face must be flat and no burrs must be produced.

Al_2O_3 Ceramic

The casings of speed reducing gearboxes are normally made from cast iron. They are completely machined in a machining centre where one of the operations involves boring the hole which carries the output shaft.

WORKPIECE	–	Reduction gearbox casing
MATERIAL	–	GG 30 grey cast iron
OPERATION	–	Boring the output shaft hole 175 mm dia × 30
MACHINE	–	Scharmann Solon 2 Machining Centre
TOOLHOLDER	–	Boring toolholder for round indexable inserts
INSERT	–	RNGN 120700 T
GRADE	–	Al_2O_3 'White' ceramic
EDGE CONDITION	–	Chamfered
CUTTING SPEED	–	488 m min^{-1}
DEPTH OF CUT	–	3 mm
FEED	–	0.25 mm/rev
COOLANT	–	No

Si Based Ceramic

One of the features of Silicon Nitride ceramic as a cutting material is its ability to machine cast iron with interrupted cutting conditions at high speeds. This example is taken from the machining of a brake disc. One of the operations is the machining of the face of the boss of the disc which has holes drilled through to take the bolts used to mount the disc. These holes create a considerable interrupted cutting situation.

WORKPIECE	– A grey cast iron brake disc
MATERIAL	– GG25
OPERATION	– Facing from 160 mm to 79 mm diameter
MACHINE	– Special purpose machine
TOOLHOLDER	– Indexable insert holder for square inserts
INSERT	– SNGN 120416T
GRADE	– Silicon nitride ceramic
EDGE CONDITION	– Chamfered
CUTTING SPEED	– 550 m min^{-1}
DEPTH OF CUT	– 2 mm
FEED	– 0.05 mm/rev
COOLANT	– No

6.4 DUCTILE CAST IRON

Examples of machining ductile cast iron are given below using Uncoated Hardmetal, Coated Hardmetal and Al_2O_3 Ceramic as the cutting material.

Uncoated Hardmetal

A flywheel for a marine engine is to be machined on the inner and outer flange faces.

WORKPIECE	– Cast iron flywheel
MATERIAL	– BS 3333 grade P 440/7
OPERATION	– Facing the inner and outer flange faces
MACHINE	– Manually operated vertical boring machine
TOOL	– 75° approach angle
RAKE ANGLE	– 5° negative
CORNER RADIUS	– 1.6 mm
GRADE	– ISO application group K20
EDGE CONDITION	– Rounded
CUTTING SPEED	– 160 – 45 m min^{-1} (constant revs.)
DEPTH OF CUT	– 3 – 4 mm
FEED	– 0.5 mm/rev
COOLANT	– No

Coated Hardmetal

The bearing diameter of a cast iron axle box casing has to be bored prior to burnishing.

WORKPIECE	–	Ductile cast iron casing
MATERIAL	–	GTW 55
OPERATION	–	Boring the bearing diameter
MACHINE	–	Scharmann machining centre
TOOL	–	90° approach angle
RAKE ANGLE	–	3° positive
CORNER RADIUS	–	0.8 mm
GRADE	–	ISO application group K15
EDGE CONDITION	–	Rounded
CUTTING SPEED	–	220 m min^{-1}
DEPTH OF CUT	–	1 mm
FEED	–	0.25 mm/rev
COOLANT	–	No

Al_2O_3 Ceramic

Ceramics are used to turn, face and taper bore the pistons which are part of the hydraulic systems in earth moving equipment. The case described is the turning of the outside diameter of a piston.

WORKPIECE	–	A cast piston body 150 mm diameter
MATERIAL	–	Ductile cast iron
OPERATION	–	Turning the diameter and face of the piston
MACHINE	–	Warner Swasey 2-SC turret lathe
TOOLHOLDER	–	Indexable insert toolholder for square inserts
APPROACH ANGLE	–	75°
INSERT	–	SNGN 120816 T
RAKE ANGLE	–	negative 6°
GRADE	–	Al_2O_3 'White' ceramic
EDGE CONDITION	–	Chamfered
CUTTING SPEED	–	500 m min^{-1}
DEPTH OF CUT	–	6 mm
FEED	–	0.32 mm/rev
COOLANT	–	No

6.5 HARD CAST IRON

For machining hard cast irons the cases which are given cover Uncoated Hardmetal, Coated Hardmetal, Al_2O_3 Ceramic and CBN.

Uncoated Hardmetal

The shoulder of a chilled cast iron roll barrel is being turned using a broad uncoated hardmetal tool. The hardness of the roll is 80 Shore. The journals of the roll are made in alloyed cast iron and the depth from the barrel to the journals is 100 mm.

WORKPIECE	– Mill Roll
MATERIAL	– Chilled cast iron 80 Shore
OPERATION	– Turning the shoulder of the roll barrel 100 mm.
MACHINE	– VDF Roll Lathe
TOOL	– 90° (plunging in)
RAKE ANGLE	– 0°
CORNER RADIUS	– 2 mm
GRADE	– ISO application group K05
EDGE CONDITION	– Negative land, 1 mm at 5°
CUTTING SPEED	– 8 m min^{-1}
DEPTH OF CUT	– 0.35 mm
FEED	– 0.35 mm/rev
COOLANT	– No

Coated Hardmetal

A roll similar to the one used as an example of machining hard cast iron with uncoated hardmetal is being profiled on the diameter.

WORKPIECE	– Chilled Cast iron roll
MATERIAL	– Chilled cast iron 80 Shore
OPERATION	– Finish turning the profile
MACHINE	– Craven lathe
TOOL	– 93° approach angle PDJNR
RAKE ANGLE	– 5° negative
CORNER RADIUS	– 1.2 mm
GRADE	– ISO application group K05-K15
EDGE CONDITION	– Rounded

CUTTING SPEED	– 20 m min^{-1}
DEPTH OF CUT	– 0.5 mm
FEED	– 0.2 mm/rev
COOLANT	– No

Al_2O_3 Ceramic

The turning of chilled cast iron rolls is an operation which has been performed for some considerable time with ceramics. This example illustrates the profiling of a chilled cast iron roll whereby a series of grooves are produced in the surface of the roll in a wave like pattern. The dimensions quoted do not include the shafts of the roll and only relate to the working surface of the roll itself.

WORKPIECE	– A roll 290 mm diameter × 500 mm long
MATERIAL	– Chilled cast iron 75 Shore hardness
OPERATION	– Profile turning the roll surface
MACHINE	– CNC Turning Machine
TOOLHOLDER	– CRDCN 2525 M12
INSERT	– RCGX 120800 T
GRADE	– Al_2O_3 'White' ceramic
EDGE CONDITION	– Chamfered
CUTTING SPEED	– 80 m min^{-1}
DEPTH OF CUT	– 0.5 mm
FEED	– 0.3 mm/rev
COOLANT	– No

CBN

This case concerns the work rolls on a steel strip mill. They are 550 mm in diameter and 1200 mm long. They are made from chilled cast iron and have a hardness of 78 – 88 Shore. When the roll is being refurbished, up to 12 mm of stock has to be removed from the diameter. Instead of the previously accepted method of grinding these rolls are now turned with CBN.

WORKPIECE	– A hard cast iron roll 550 mm dia. × 1200 mm
MATERIAL	– Chilled cast iron, hardness 78 – 88 Shore
OPERATION	– Removing 12 mm from the diameter by turning

MACHINE	– Craven centre lathe
TOOLHOLDER	– Indexable insert holder with -6° rake
INSERT	– 12.7 mm diameter, round insert
GRADE	– CBN
EDGE CONDITION	– Chamfered
CUTTING SPEED	– 46 m min^{-1}
DEPTH OF CUT	– 0.1 mm
FEED	– 0.6 mm/rev
COOLANT	– No

6.6 FREE MACHINING & LOW C STEELS

The examples of cutting materials used to illustrate the machining of this group of steels are High Speed Steel, Uncoated Hardmetal and Coated Hardmetal.

High Speed Steel

A threaded collar is being machined from a mild steel bar. The bar is turned on the diameter then drilled, threaded in the bore and finally parted off. The operation described is the turning of the outside diameter.

The bar is nominally 100 mm diameter. The final O.D. of the collar is 87.3 mm and this is produced in two machining passes. The length of bar turned down to the collar diameter is 35 mm.

WORKPIECE	– Mild steel bar 100 mm diameter × 350 mm long
MATERIAL	– En 8
OPERATION	– Turning the diameter to 87.3 mm in 2 passes
MACHINE	– Colchester lathe
TOOL	– High Speed Steel tool bit ground to suit
RAKE ANGLE	– + 12°
CORNER RADIUS	– 0.8 mm
EDGE CONDITION	– Sharp
CUTTING SPEED	– 35 m min^{-1} 1st pass
	– 50 m min^{-1} 2nd pass
DEPTH OF CUT	– 4 mm 1st pass
	– 2.3 mm 2nd pass

FEED	– 0.6 mm/rev 1st pass
	– 0.2 mm/rev 2nd pass
COOLANT	– Yes

Uncoated Hardmetal

A shaft has a screw thread machined on it and then requires an undercut to be produced at the end of the thread.

WORKPIECE	– A threaded shaft
MATERIAL	– C 35
OPERATION	– Turning an undercut
MACHINE	– Centre lathe
TOOL	– Special brazed hardmetal tool
RAKE ANGLE	– 8° positive
CORNER RADIUS	– 2 mm
GRADE	– ISO application group P15
EDGE CONDITION	– Lightly honed
CUTTING SPEED	– 130 m min^{-1}
DEPTH OF CUT	– 4 mm
FEED	– 0.12 mm/rev
COOLANT	– No

Coated Hardmetal

The case chosen here is that of turning the outside diameter of an adaptor.

WORKPIECE	– An adaptor
MATERIAL	– Ck 15
OPERATION	– Turning the body diameter
MACHINE	– CNC lathe
TOOL	– 95° approach angle PCLNR
RAKE ANGLE	– 5° positive
CORNER RADIUS	– 0.8 mm
GRADE	– ISO application group P05-P20
EDGE CONDITION	– Rounded
CUTTING SPEED	– 280 m min^{-1}
DEPTH OF CUT	– 2 – 3 mm
FEED	– 0.25 mm/rev
COOLANT	– No

6.7 ALLOY & MEDIUM TO HIGH C STEELS

Only two cutting materials have been chosen as examples to use for the machining of Alloy and Medium to High C Steels and these are Uncoated Hardmetal and Coated Hardmetal.

Uncoated Hardmetal

The rough forged shaft which is taken for this example is to be turned down to the several diameters required prior to the finishing operation. There is a considerable amount of interrupted cutting and this is a good example of heavy machining.

WORKPIECE	– A rough forged shaft
MATERIAL	– 32CrMoV12 10
OPERATION	– Turning the shaft diameters
MACHINE	– Centre lathe
TOOL	– 75° approach angle
RAKE ANGLE	– 5° negative
CORNER RADIUS	– 1.6 mm
GRADE	– ISO application group P40
EDGE CONDITION	– Chamfered
CUTTING SPEED	– 85 – 120 m min^{-1} (constant revs)
DEPTH OF CUT	– 4 – 5 mm
FEED	– 0.45 mm/rev
COOLANT	– No

Coated Hardmetal

Crankshaft turning is also carried out in addition to crankshaft milling. In this case the bearings of the crankshaft are being turned with coated hardmetal.

WORKPIECE	– Forged crankshaft
MATERIAL	– St E 43
OPERATION	– Turning the bearings
MACHINE	– Swedturn CNC lathe
TOOL	– Round indexable insert toolholder
RAKE ANGLE	– 0°
CORNER RADIUS	– 16 mm diameter

GRADE	– ISO application group P35
EDGE CONDITION	– Chamfered
CUTTING SPEED	– 140 m min^{-1}
DEPTH OF CUT	– 2 – 4 mm
FEED	– 0.5 – 1.2 mm/rev
COOLANT	– No

6.8 TOOL STEELS

For machining Tool Steels in the soft condition Uncoated Hardmetal and Coated Hardmetal are chosen for the examples of cutting materials used. When machining Hard Tool Steels Al_2O_3 Ceramic and CBN are added to the above cutting material examples.

6.8.1 SOFT CONDITION

Uncoated Hardmetal

Reamers are turned from bar in the soft condition prior to hardening and final grinding. They are then parted off. This example details the cutting conditions for the parting off operation.

WORKPIECE	– Tool steel bar in the soft condition
MATERIAL	– M2 HSS in the soft condition
OPERATION	– Parting off
MACHINE	– NC lathe
TOOL	– Brazed hardmetal parting off tool
RAKE ANGLE	– 0°
GRADE	– ISO application group P40
EDGE CONDITION	– Lightly honed
CUTTING SPEED	– 80 m min^{-1}
WIDTH OF CUT	– 4 mm
FEED	– 0.17 – 0.05 mm/rev
COOLANT	– Yes

The reamer used as the example is a long series reamer and is unsupported during the parting off operation thus the feed is reduced when the diameter reaches 10 mm.

Coated Hardmetal

This is an example of a high speed steel end mill blank being turned on the diameter where the flutes will be cut and facing the end of the cutter blank in the same operation.

WORKPIECE	– High speed steel end mill blank
MATERIAL	– M2 HSS in the soft condition
OPERATION	– Turning the flute diameter and facing the end
MACHINE	– CNC lathe
TOOL	– 95° approach angle
RAKE ANGLE	– 0°
CORNER RADIUS	– 1.2 mm
GRADE	– ISO application group P25
EDGE CONDITION	– Rounded
CUTTING SPEED	– 150 m min^{-1}
DEPTH OF CUT	– 2 mm
FEED	– 0.3 mm/rev
COOLANT	– No

In this case good swarf control is important and an appropriate chip control groove must be used.

6.8.2 HARD CONDITION

Uncoated Hardmetal

A high speed steel punch is being machined on the diameter in this example.

WORKPIECE	– High speed steel punch
MATERIAL	– M2 HSS hardened to 62 Rc
OPERATION	– Turning the diameter of the punch
MACHINE	– Centre lathe
TOOL	– 93° approach angle
RAKE ANGLE	– 7° positive
CORNER RADIUS	– 0.8 mm
GRADE	– ISO application group K10
EDGE CONDITION	– Rounded
CUTTING SPEED	– 28 m min^{-1}

DEPTH OF CUT	– 0.5 – 1 mm
FEED	– 0.1 mm/rev
COOLANT	– No

Coated Hardmetal

A similar punch used as the example above is being turned on another machine with coated hardmetal.

WORKPIECE	– High speed steel punch
MATERIAL	– M2 HSS hardened to 62 Rc
OPERATION	– Turning the diameter of the punch
MACHINE	– NC lathe
TOOL	– 95° approach angle
RAKE ANGLE	– 7° positive
CORNER RADIUS	– 0.8 mm
GRADE	– ISO application group K20/P25
EDGE CONDITION	– Rounded
CUTTING SPEED	– 52 m min^{-1}
DEPTH OF CUT	– 0.8 mm
FEED	– 0.12 mm/rev
COOLANT	– No

Al_2O_3 Ceramic

In this example punches which are to be used for cold heading applications are being turned in the hard condition prior to final finishing by grinding. They are made from M2 High Speed Steel.

WORKPIECE	– Hardened blank 25 mm Diameter × 120 mm long
MATERIAL	– M2 HSS hardened to 62 Rc
OPERATION	– Turning the diameter over the whole length
MACHINE	– Centre lathe
TOOLHOLDER	– Indexable insert toolholder for round inserts
INSERT	– 12.7 mm round insert
RAKE ANGLE	– Negative 6°
GRADE	– Al_2O_3 'Black' Ceramic (containing TiC)

EDGE CONDITION – Chamfered
CUTTING SPEED – 100 m min^{-1}
DEPTH OF CUT – 0.4 mm
FEED – 0.15 mm/rev
COOLANT – No

The black ceramic is recommended for machining steels having a hardness above 46 Rc and up to 65 Rc. However, for steels of 60 Rc and above the feed must be less than the width of the chamfer applied at the cutting edge. This chamfer is usually of the order of 0.2 mm at an angle of 20°.

CBN

Thread rolling dies are used in pairs of rolls made from hardened M2 high speed steel. They are reground after they are worn and are then recut to provide a new thread. Each roll can take about one hour to grind away the old grooves. By turning with CBN the time can be cut to around five minutes.

WORKPIECE – A worn thread rolling die 100 mm O.D. × 50 mm
MATERIAL – M2 high speed steel hardened to 63 – 64 Rc
OPERATION – Turning the O.D.
MACHINE – Herbert No.8 Preoptive turret lathe
TOOLHOLDER – Indexable insert holder with -6° rake
INSERT – 12.7 mm diameter, round insert
GRADE – CBN
EDGE CONDITION – Chamfered
CUTTING SPEED – 95 m min^{-1}
DEPTH OF CUT – up to 2
FEED – 0.1 mm/rev
COOLANT – No

6.9 AUSTENITIC STAINLESS STEELS

Examples of cutting materials chosen to illustrate the machining of Austenitic Stainless Steels are Uncoated Hardmetal, Coated Hardmetal and Cermet.

Uncoated Hardmetal

In this case the workpiece is a large valve body made from cast austenitic stainless steel. Stability is very poor and the operation is prone to the setting up of vibration. The flange on the body is being turned and faced.

WORKPIECE	– Austenitic stainless steel valve body
MATERIAL	– EN 56A
OPERATION	– Turning and facing the flange
MACHINE	– Manually operated vertical boring machine
TOOL	– 45° approach angle PSSNR
RAKE ANGLE	– 5° negative
CORNER RADIUS	– 1.6 mm
GRADE	– ISO application group K30
EDGE CONDITION	– Chamfered
CUTTING SPEED	– 70 m min^{-1}
DEPTH OF CUT	– 2 – 5 mm
FEED	– 0.5 mm/rev
COOLANT	– No

Coated Hardmetal

A workpiece connector is being machined from bar. This case is the turning of the diameter of the connector.

WORKPIECE	– Austenitic stainless steel bar
MATERIAL	– EN 58M
OPERATION	– Turning down the diameter to finished size
MACHINE	– CTC 4 NC
TOOL	– 95° approach angle
RAKE ANGLE	– 5° negative
CORNER RADIUS	– 0.8 mm
GRADE	– ISO application group K25/P25
EDGE CONDITION	– Rounded
CUTTING SPEED	– 200 m min^{-1}
DEPTH OF CUT	– 2 mm
FEED	– 0.3 mm/rev
COOLANT	– Yes

Cermet

Cermets perform well in machining stainless steels where light cuts are involved and retaining the sharpness of the cutting edge is vital to the success of the operation.

Threading is typical of this type of cutting condition and the example quoted is one where an internal thread is being machined in a stainless steel bush.

WORKPIECE	– A stainless steel bush
MATERIAL	– X10CrNi18.9
OPERATION	– Machining an internal thread M34 × 1.5
MACHINE	– Lathe
TOOLHOLDER	– Internal threading holder
INSERT	– TPMX 1603 IR 150 M Internal threading insert
GRADE	– Cermet
EDGE CONDITION	– Sharp
PASSES	– 6
CUTTING SPEED	– 100 m min^{-1}
COOLANT	– Yes

Although coolant can cause thermal shock, in this case the cut is so light that there is no problem thus the coolant is helpful in removing the chips from the bore.

6.10 FERRITIC & MARTENSITIC STAINLESS STEELS

Two cutting materials are used to illustrate the machining of these stainless steels. They are Uncoated Hardmetal and Coated Hardmetal.

Uncoated Hardmetal

In this example it is required to turn a bar of stainless steel to produce the profile of a shaft 22 mm diameter × 440 mm long.

WORKPIECE	– Stainless steel bar
MATERIAL	– EN 57 annealed condition
OPERATION	– Turning the profile of the shaft
MACHINE	– CNC lathe

TOOL	– SDJCR 2020 11
INSERT	– DCMT 110304
GRADE	– ISO application group P30
EDGE CONDITION	– Rounded
CUTTING SPEED	– 95 m min^{-1}
DEPTH OF CUT	– 0.3 mm
FEED	– 0.5 mm/rev
COOLANT	– Yes

Coated Hardmetal

This example is the turning of the boss of a stainless steel ring. The boss diameter is 89 mm and its length is 35 mm.

WORKPIECE	– Stainless steel ring
MATERIAL	– AISI 416
OPERATION	– Turning a boss 89 mm dia. × 35 mm long
MACHINE	– Okuma LC 20 CNC lathe
TOOL	– 93° approach angle
INSERT	– CNMG 120408 with appropriate chipgroove
GRADE	– ISO application group P25
EDGE CONDITION	– Rounded
CUTTING SPEED	– 105 m min^{-1}
DEPTH OF CUT	– 2.5 mm
FEED	– 0.25 mm/rev
COOLANT	– Yes

6.11 HEAT RESISTING ALLOYS

For this very difficult group of workpiece materials Uncoated Hardmetal, Coated Hardmetal, and Silicon Based Ceramic are chosen as cutting materials for the machining examples.

Uncoated Hardmetal

A disc cast in a cobalt chromium heat resisting alloy is being machined. The turning and facing of the outer part of the disc is being done with uncoated hardmetal.

WORKPIECE	– Cobalt alloy disc
MATERIAL	– 65% Co 27% Cr
OPERATION	– Turning and facing the outer part of the disc
MACHINE	– CNC lathe
TOOL	– PCLNR 2525
INSERT	– CNMG 120408 with appropriate chipgroove
GRADE	– ISO application group K20
EDGE CONDITION	– Rounded
CUTTING SPEED	– 65 m min^{-1}
DEPTH OF CUT	– 1 mm
FEED	– 0.07 mm/rev
COOLANT	– Yes, good supply preferred

Coated Hardmetal

Bars of Nimonic 600 are rough machined prior to being used as electrodes in the melting furnace.

WORKPIECE	– A rough forged bar
MATERIAL	– Nimonic 600
OPERATION	– Rough turning the diameter
MACHINE	– Centre lathe (60 HP)
TOOL	– PSBNR 4040
INSERT	– SNMM 190624 with appropriate chipgroove
GRADE	– ISO application group K25-K35 multi-layer coated
EDGE CONDITION	– Rounded
CUTTING SPEED	– 55 m min^{-1}
DEPTH OF CUT	– 3 – 4 mm
FEED	– 0.6 mm/rev
COOLANT	– Yes, copious supply needed

In this case the chipgroove is very important. Although the insert is inclined negatively in the tool, the chipgroove has an effective 6° positive rake built in. The chipgroove is also designed for roughing operations.

Si BASED CERAMIC

Two examples of machining with silicon based ceramics are given. In the first case a sialon is being used to machine a forged Inconel ring. A round sialon insert is turning a profile on the face of the ring.

The second example is a Waspaloy turbine disc which is being turned on the outer part of the face and where the cutting material is alumina reinforced with silicon carbide whiskers.

Example 1

WORKPIECE	– A forged ring approximately 600 mm diameter
MATERIAL	– Inconel 718
OPERATION	– Profiling the outer face from 427 mm diameter
MACHINE	– Not given
TOOLHOLDER	– MRGNR 3225 P12
INSERT	– RNGN 120700T
GRADE	– Sialon
EDGE CONDITION	– Chamfered
CUTTING SPEED	– 305 m min^{-1}
DEPTH OF CUT	– 0.51 mm
FEED	– 0.2 mm/rev
COOLANT	– Yes

Example 2

WORKPIECE	– A turbine disc 600 mm diameter
MATERIAL	– Waspaloy
OPERATION	– Turning the outer face
MACHINE	– Not given
TOOLHOLDER	– MRGNL 3225 P12
INSERT	– RNGN 120700T
GRADE	– Al_2O_3 – silicon carbide whisker reinforced
EDGE CONDITION	– Chamfered
CUTTING SPEED	– 213 m min^{-1}
DEPTH OF CUT	– 2.54 mm
FEED	– 0.18 mm/rev
COOLANT	– Yes

6.12 TITANIUM & Ti ALLOYS

In the case of Titanium and Ti Alloys only one example is quoted and that is Uncoated Hardmetal.

Uncoated Hardmetal

For an example of machining a titanium alloy the operation chosen is the drilling of a hole in a titanium alloy bar to produce a bush. The hole is a blind hole and the bush is then parted off and redrilled afterwards.

WORKPIECE	– Titanium alloy bar
MATERIAL	– Ti-6Al-4V
OPERATION	– Drilling an 18 mm diameter hole 40 mm deep
MACHINE	– Okuma NC
TOOL	– 18 mm diameter short hole drill
RAKE ANGLE	– 5° positive
CORNER RADIUS	– 0.4 mm
GRADE	– ISO application group K20
EDGE CONDITION	– 0.02 radius
CUTTING SPEED	– 35 m min^{-1}
FEED	– 0.05 mm/rev
COOLANT	– Yes, copious supply needed

6.13 PLASTICS & NON-METALLICS

The cutting materials chosen as examples for machining these materials are Uncoated Hardmetal and PCD.

Uncoated Hardmetal

It is required to machine a guide bush from a 'Tufnel' bar. The example quoted is the turning of the outside diameter of the guide bush.

WORKPIECE	– A Tufnel bar
MATERIAL	– Tufnel
OPERATION	– Finish turning the O.D.
MACHINE	– Frontier lathe
TOOL	– 93° approach angle
RAKE ANGLE	– High positive, Aluminium geometry
CORNER RADIUS	– 0.8 mm
GRADE	– ISO application group K10
EDGE CONDITION	– Sharp
CUTTING SPEED	– 800 m min^{-1}
DEPTH OF CUT	– 1 mm

FEED	– 0.3 mm/rev
COOLANT	– No

PCD

One of the hardest forms of granite which can be found is used as the lapping wheel on certain lapping machines. In this case the granite plates are 350 mm diameter × 75 mm thick and they have a bore at the centre which is used to mount the wheel.

The face has to be machined flat and PCD can do this operation in one pass without losing size.

WORKPIECE	– Lapping wheel 350 mm O.D. × 70 mm I.D. × 75 mm
MATERIAL	– Extremely hard granite
OPERATION	– Turning the face of the wheel
MACHINE	– Centre Lathe
TOOLHOLDER	– Special holder with -30° rake
INSERT	– Special insert with 6.3 mm radius
GRADE	– PCD
EDGE CONDITION	– No special edge reinforcement
CUTTING SPEED	– 48 m min^{-1}
DEPTH OF CUT	– 3 mm
FEED	– 0.4 mm/rev
COOLANT	– Yes

6.14 HARDMETAL

Hardmetal dies, rolls and other forming tools are almost always ground, using diamond wheels, to bring them to the required dimensions. Many of the parts involved are round and so lend themselves to the possibility of turning if a suitable cutting material is used.

Because they have extremely high hardness both CBN and PCD can be used to turn the Co-WC hardmetals. However, it is not practical to use CBN on the lower cobalt content hardmetals.

CBN

Hardmetal rolls are used in the finishing section of a rod rolling mill. The last 10 pairs of rolls are made from Co-WC grades of hardmetal.

One of the roll sizes is 156 mm O.D. × 92 mm I.D. × 61 mm wide. The sides and outside diameter are turned to finished dimensions after sintering. The bore is ground to achieve the high dimensional tolerance necessary. The groove profile of the roll is ground by the customer as required.

WORKPIECE	– A hardmetal Morgan roll
MATERIAL	– 15% Co 85% WC hardmetal 1000 VDH
OPERATION	– Turning the O.D. and facing each side
MACHINE	– Lathe
TOOLHOLDER	– Indexable insert holder with -6° rake
INSERT	– 12.7 mm diameter, round insert
GRADE	– CBN
EDGE CONDITION	– Chamfered
CUTTING SPEED	– 14 m min^{-1}
DEPTH OF CUT	– 0.1 mm
FEED	– 0.08 mm/rev
COOLANT	– No

No coolant is used. When this operation is done by grinding it takes about four times as long.

PCD

A hardmetal bar drawing die pellet is usually ground before being fitted into a steel supporting case. PCD is now being used to turn the outside diameter instead of grinding.

WORKPIECE	– A hardmetal hollow cylinder 45 mm O.D. × 25 mm
MATERIAL	– 12% Co 88% WC hardmetal 1200 VDH
OPERATION	– Turning the O.D.
MACHINE	– Lathe
TOOLHOLDER	– Special insert toolholder with -6° rake
INSERT	– Small round solid PCD insert
GRADE	– PCD
EDGE CONDITION	– Sharp
CUTTING SPEED	– 16 m min^{-1}
DEPTH OF CUT	– 0.1 mm
FEED	– 0.3 mm/rev
COOLANT	– No

Hardmaterial Cutting Tool and Associated Standards

	ISO Reference	BSI Reference
Brazed Turning Tools and Tips:		
Dimensions of shanks	ISO/241	
Dimensions of tips	ISO/242	
Dimensions of external tools	ISO/243	
Designation and marking	ISO/504	
Dimensions of internal tools	ISO/514	
Indexable Inserts:		
Dimensions without fixing holes	ISO/883	BS4193 Pt2
Designation	ISO/1832	BS4193 Pt1
Designation, letter symbols for double chamfers	ISO/1832DAD1	
Dimensions with cylindrical fixing holes	ISO/3364	BS4193 Pt3
Dimensions for milling	ISO/3365	BS4193 Pt15
Dimensions for milling, Wiper	ISO/3365 Pt3	
Dimensions with part cylindrical fixing holes, 7cl	ISO/6987 Pt1	BS4193 Pt13
Dimensions with part cylindrical fixing holes, 11cl	ISO/6987 Pt2	
Dimensions of inserts, style 35' vee	ISO/TR6987 Pt3	
Dimensions of ceramic, without fixing hole	ISO/9361 Pt1	
Dimensions of ceramic, with cylindrical fixing holes	ISO/9361 Pt2	
Chip control ranges	ISO/CD11910	

Toolholders for Indexable Inserts:		
Designation of toolholders and cartridges	ISO/5608	BS4193 Pt6
Designation of toolholders and cartridges, style H	ISO/5608DAD1	
Designation of toolholders and cartridges, style P	ISO/5068/AM1	
Dimensions of boring bars	ISO/5609	BS4193 Pt18
Dimensions of boring bars, style Q	ISO/5609DAD1	
Dimensions of toolholders, turning and copying	ISO/5610	BS4193 Pt7
Dimensions of toolholders, small tools	ISO/5610DAD1	
Dimensions of toolholders, style H	ISO/5610DAD2	
Dimensions of cartridges, type A	ISO/5611	BS4193 Pt8
Dimensions of cartridges, h=8	ISO/5611 Addendum 1	
Dimensions of cartridges, h=6	ISO/5611DAD2	
Designation of boring bars	ISO/6261	BS4193 Pt14
Designation of boring bars, style P (117.5 deg.)	ISO/6261/AM1	
Milling Cutters for Indexable Inserts:		
Dimensions of end mills, parallel shanks	ISO/6262 Pt1	BS4193 Pt9
Dimensions of end mills, Morse taper shanks	ISO/6262 Pt2	BS4193 Pt10
Dimensions of face mills	ISO/6462	BS4193 Pt11
Dimensions of side and face mills	ISO/6986	BS4193 Pt12
Designation of bore type cutters	ISO/7406	BS4193 Pt17
Designation of shank type cutters	ISO/7848	BS4193 Pt16
Milling Cutters other than for Inserts:		
Dimensions of plain parallel shanks	ISO/3338 Pt1	BS122 Pt3
Dimensions of flatted parallel shanks	ISO/3338 Pt2	
Brazed helical end mills, parallel shanks	ISO/DP10145 Pt1	
Brazed helical end mills, 7/24 taper shanks	ISO/DP10145 Pt2	
Dimensions of solid hardmetal end mills	ISO/CD10911	

Designation of solid hardmetal end mills	ISO/CD11529

Other Standards:

Groups of applications for cutting tools	ISO/513:1975
Groups of applications for cutting tools, amended	ISO/DIS/513
Geometry of cutting tools, general terms	ISO/3002 Pt1
Geometry of cutting tools, conversion formulae	ISO/3002 Pt2
Geometry of cutting tools, quantities in cutting	ISO/3002 Pt3
Geometry of cutting tools, forces, energy and power	ISO/3002 Pt4
Corner radii for single point cutting tools	ISO/3286
Tool life testing, single point tools	ISO/6385
Carbide burrs	ISO/7755 Pt1 to Pt12
Tool life testing, face milling	ISO/8688 Pt1
Tool life testing, end milling	ISO/8688 Pt2
Dimensions of flatted parallel shanks for drills	ISO/DIS9766

Further Information and Reading

Most of the cutting tool manufacturers include excellent technical information in the variety of brochures which they offer and much of this material is very detailed and helpful. In addition, if one is considering a new project or having machining difficulties with a component or workpiece material then these tool manufacturers usually offer the services of competent cutting tool specialists who have a wealth of experience at hand.

Manufacturers' computer software and access to cutting data banks are further possibilities which should not be overlooked.

One excellent publication provides information on the chemical compositions and mechanical properties of almost all the manufactured hard cutting materials existing in the world (uncoated and coated hardmetal, cermets, ceramics, CBN and PCD). It also includes good information on hardmetal manufacture etc. This book is entitled *The world Directory and Handbook of Hardmetals* by Kenneth J.A. Brookes.

Index

Recent publications from THE INSTITUTE OF MATERIALS

OF